Handbook of CDMA System Design, Engineering, and Optimization

Prentice Hall Communications Engineering and Emerging Technologies Series

Theodore S. Rappaport, Series Editor

KIM *Handbook of CDMA System Design, Engineering, and Optimization*

GARG *IS-95 CDMA and cdma2000: Cellular/PCS Systems Implementation*

GARG & WILKES *Principles and Applications of GSM*

HAĆ *Multimedia Applications Support for Wireless ATM Networks*

LIBERTI & RAPPAPORT *Smart Antennas for Wireless Communications: IS-95 and Third Generation CDMA Applications*

RAPPAPORT *Wireless Communications: Principles & Practice*

RAZAVI *RF Microelectronics*

STARR, CIOFFI & SILVERMAN *Understanding Digital Subscriber Line Technology*

FORTHCOMING

CIMINI & LI *Orthogonal Frequency Division Multiplexing for Wireless Communication*

LO & ABU-DAYYA *Fixed Wireless Communication*

MOLISCH *Wideband Wireless Digital Communication*

POOR & WANG *Wireless Communication Systems: Advanced Techniques for Signal Reception*

REED & WOERNER *Software Radio: A Modern Approach to Radio Engineering*

TRANTER, KOSBAR, RAPPAPORT & SHANMUGAN *Simulation of Modern Communications Systems with Wireless Applications*

Handbook of CDMA System Design, Engineering, and Optimization

The Members of Technical Staff, Bell Labs
Edited by Kyoung Il Kim

Prentice Hall PTR
Upper Saddle River, NJ 07458
www.phptr.com

Library of Congress Cataloging-in-Publication Data

```
Handbook of CDMA system design, engineering and optimization / edited by Kyoung Il Kim.
    p. cm.
  Includes bibliographical references and index.
  ISBN 0-13-017572-2
  1. Code division multiple access. 2. Mobile communications systems. I. Kim, Kyoung
Il. II. Title.

TK5103.452 .H36 1999
621.3845--dc21
                                                                          99-055012
```

Editorial/Production Supervision: *Joan L. McNamara*
Acquisitions Editor: *Bernard Goodwin*
Marketing Manager: *Lisa Konzelmann*
Editorial Assistant: *Diane Spina*
Cover Design Director: *Jerry Votta*
Cover Designer/Illustration: *Marjorie Paradise-Persiano, Advanced Media Solutions, Lucent Technologies*
Manufacturing Manager: *Alexis R. Heydt*
Composition: *Aurelia Scharnhorst*

©2000 by Prentice Hall PTR
Prentice-Hall, Inc.
Upper Saddle River, New Jersey 07458

Prentice Hall books are widely used by corporations and government agencies for training, marketing, and resale.
The publisher offers discounts on this book when ordered in bulk quantities. For more information, contact Corporate Sales Department. Phone: 800-382-3419; Fax: 201-236-7141; E-mail: corpsales@prenhall.com; Or write: Prentice Hall PTR, Corp. Sales Dept., One Lake Street, Upper Saddle River, NJ 07458

Product names mentioned herein are the trademarks of their respective owners.

All rights reserved. No part of this book may be reproduced, in any form or by any means, without permission in writing from the publisher.

Printed in the United States of America
    10 9 8 7 6 5 4 3 2 1

ISBN 0-13-017572-2

Prentice-Hall International (UK) Limited, *London*
Prentice-Hall of Australia Pty. Limited, *Sydney*
Prentice-Hall Canada Inc., *Toronto*
Prentice-Hall Hispanoamericana, S.A., *Mexico*
Prentice-Hall of India Private Limited, *New Delhi*
Prentice-Hall of Japan, Inc., *Tokyo*
Pearson Education Asia Pte. Ltd.
Editora Prentice-Hall do Brasil, Ltda., *Rio de Janeiro*

Table of Contents

Preface	xi
Foreward	xv

1 Introduction	**1**
1.1 References	4
2 CDMA Overview	**7**
2.1 Concept	7
2.2 Attributes	8
2.2.1 Capacity	8
2.2.2 Power Control	10
2.2.3 Soft Handoff	11
2.2.4 Voice Activity	11
2.3 Reference	12
3 Spectrum Coordination	**13**
3.1 Spectrum Coordination	13
3.1.1 Spectrum Allocation and Channel Numbering	14
3.1.2 Channel Availability	15

	3.1.3 Preferred Channels	16
	3.1.4 Intra-System Frequency Planning Issues	17
	3.1.5 Inter-System Frequency Planning Issues	18
3.2	References	22

4 Pilot Assignment 23

4.1	Introduction	23
4.2	Pilot Channel Model	24
4.3	Call Processing for a Mobile Station	26
4.4	PILOT_INC and Pilot Phase Offset Reuse	28
	4.4.1 Pilot Separation and a Lower Limit for PILOT_INC	29
	4.4.2 Reusing Pilot PN Sequence Phase Offsets	34
	4.4.3 An Upper Limit for PILOT_INC	41
4.5	A Procedure for Phase Assignment	41
4.6	References	45

5 Mobile Station Access and Paging 47

5.1	Description of Mobile Station Access Protocol	47
5.2	Average Persistence Delay for Access Request Attempt	54
5.3	Access Channel Capacity	58
	5.3.1 Access Channel Simulation Model	59
	5.3.2 Access Channel Simulation Results	61
	5.3.3 Access Channel Capacity Analysis	62
5.4	Paging Channel Capacity	66
	5.4.1 Paging Channel Characteristics	67
	5.4.2 Assumptions	70
	5.4.3 Paging Channel Capacity	72
	5.4.4 Summary	80
5.5	References	80

6 Handoff 83

6.1	Hard Handoff	83
6.2	Soft and Softer Handoff	84
	6.2.1 Definition	84
	6.2.2 Procedure	84

6.2.3 Comparisons	86
6.2.4 Performance	87
6.2.5 Parameters	97
6.3 Inter-Carrier Handoffs	101
6.3.1 Pocketed System	101
6.3.2 Disjoint System	104
6.4 References	107

7 Link Budgets — 109

7.1 Derivation of the Reverse-Link Budget	109
7.1.1 Required Power at the Input of the Base Station Antenna	110
7.1.2 Fading Margin	114
7.1.3 Soft Handoff Gain	117
7.1.4 Maximum Supportable Over-the-Air Path Loss	119
7.2 Forward Link	122
7.3 References	125

8 Capacity — 127

8.1 Reverse-Link Capacity	128
8.2 Forward-Link Capacity	133
8.3 Reference	137

9 Coverage — 139

9.1 Reverse-Link Coverage Area	139
9.2 Forward-Link Coverage Area	143
9.2.1 Coverage Probability for Pilot Channel	143
9.3 A Derivation of Coverage Probability for Pilot Channel	160
9.4 Reference	164

10 Traffic Engineering — 165

10.1 Introduction	165
10.2 Analysis	166
10.2.1 Two-Sector Case	166
10.2.2 Three-Sector Case	169
10.2.3 K-Sector Case	170

10.3 Numerical Results	173
10.4 References	173

11 Antennas — 175

11.1 Introduction	175
11.2 Antenna Concepts	176
11.3 Antenna System with Interference and Cell Coverage	177
11.3.1 Directional Antenna and Sectorization Gain	177
11.3.2 Coverage with Antenna Height and Gain	179
11.3.3 Reducing Interference Using Antenna Downtilt	181
11.4 Diversity Antenna Systems	183
11.4.1 Space Diversity Antenna	183
11.4.2 Polarization Diversity Antenna	184
11.4.3 Comparison of Polarization Diversity and Space Diversity	187
11.5 Antenna Isolation Guidelines for Collocated RF Stations	188
11.5.1 Introduction	188
11.5.2 Mathematical Model for Mutual Interference Evaluation	188
11.5.3 Antenna Isolation Criteria and Safe Antenna Isolation	193
11.5.4 Antenna Separation between Two Collocated RF Stations	195
11.5.5 Mutual Interference Between Multiple Collocated RF Stations	197
11.5.6 Site Survey	199
11.6 References	203

Appendix A RF Design Process — 205

A.1 Purpose	205
A.2 Process Overview	205
A.3 Preliminary Design Phase	206
A.3.1 Step 1. Project Plan and Requirements Review	206
A.3.2 Step 2. Data Preparation	208
A.3.3 Step 3. Area Visits	208
A.3.4 Step 4. Morphology Definition, Drive Test, and Calibration	209
A.3.5 Step 5. Verify Input Parameters for Coverage Prediction	210
A.3.6 Step 6. Capacity Planning	211
A.3.7 Step 7. (Iterative) Engineering	212
A.3.8 Step 8. Determine Search Rings	212

    A.3.9 Step 9. Preliminary Design Review 213
  A.4 Final Design Phase 213
    A.4.1 Step 1. Candidate-Site Selection 213
    A.4.2 Step 2. Preliminary Evaluation of Candidate-Site Coverage 215
    A.4.3 Step 3. Drive Test of Candidate Sites 216
    A.4.4 Step 4. Drive-Test Data Analysis 216
    A.4.5 Step 5. Update Parameters Needed for Coverage Prediction 217
    A.4.6 Step 6. Capacity Planning/Traffic Studies 218
    A.4.7 Step 7. Final RF Candidate Site Selection and Sketch Preparation 218
    A.4.8 Step 8. Create Input to Cell Equipment Lists 221
    A.4.9 Step 9. PN Planning 221
    A.4.10 Step 10. Create Coverage Prediction Plots 221
    A.4.11 Step 11. Design Review 221

Appendix B   Outline of RF Optimization Procedures   223

  B.1 Cluster Testing 224
    B.1.1 Spectrum Monitoring 224
    B.1.2 Basic Call Processing Test 225
    B.1.3 Unloaded Pilot Survey 225
    B.1.4 Loaded Coverage Test 226
  B.2 System-Wide Optimization 226

Appendix C   RF Coverage Prediction with CE4   229

  C.1 Overview of the CE4 Cellular Engineering Tool 229
    C.1.1 Tool Input and Scenario Setup 229
    C.1.2 Cell-Site Placement and Provisioning 230
  C.2 Analysis Features Available for the Demo Version of CE4 230
    C.2.1 Signal Strength Analysis 231
    C.2.2 CDMA Analysis 231
  C.3 System Requirements and Installation 232
  C.4 Limitations of the Demo Version of CE4 233
  C.5 Walk-Through of the CE4 Tool 234

Index   239

Preface

Until only eight years ago, the wireless telecommunications industry considered code division multiple access (CDMA) a controversial technology. However, since the Telecommunications Industry Association (TIA) published the IS-95 standard in 1993, CDMA has rapidly become a preferred choice of the industry. As of this writing, hundreds of CDMA mobile communications networks have been deployed worldwide, providing services to over 30 million subscribers. In addition, many of the radio transmission technology (RTT) proposals for the next-generation IMT-2000 system have chosen CDMA for their air interface technology.

To arrive at this level of success in the market, the CDMA standard has been revised more than once, and the in-service CDMA systems have incorporated many improvements. However, techniques for designing and tuning a system to achieve optimum performance remain to be worked out through individual research and field tests. We have attempted, in this book, to archive our findings, hoping to provide a single source of useful information for those who would otherwise have to wade through all the available literature on the topic and/or develop their own methodology. This book is the result of the recent efforts of many members of the technical staff of Bell Labs at Lucent Technologies, Inc., on CDMA technology as it applies to cellular or personal communications services (PCS). As we are now moving toward the next-generation systems, this is a good time to properly summarize all the work that has been developed and applied to the design and deployment of IS-95 CDMA systems.

In this book, we also provide an in-depth discussion of various engineering guidelines for a multitude of real-world issues involved in designing, deploying, and optimizing IS-95 CDMA systems and networks. However, because there are numerous good books on the market that provide thorough mathematics of CDMA technology, our approach has been to focus on intuitive and qualitative approaches for the various topics we discuss in this book. Nonetheless, some background in calculus and linear algebra along with probability theory will be necessary to fully understand the discussions in Chapters 5, 9, and 10.

The goal we have in mind for this book is to help readers understand (i) the types of issues involved in engineering successful CDMA mobile communication systems and (ii) the guidelines needed to deal with those issues. To make this book more useful for those who are interested in *cell planning*, we have included a demonstration version of Cellular Engineering 4 (CE4), which is an example of a CDMA coverage prediction software tool developed at Bell Labs. Using this software, readers can obtain hands-on experience in CDMA cellular engineering technique as well as in the design and analysis of CDMA networks.

Although we have attempted to cover as many relevant topics as possible, this book cannot possibly substitute for all the existing literature published on the topic of CDMA technology. Still, we believe that the various analyses and simulation results for the topics discussed will be useful for anyone seeking practical guidelines for the engineering and optimization of CDMA networks.

One last note, we have used the North American PCS CDMA system operating in the 1800 MHz spectral band as our example for numerical evaluations in many of our discussions; however, almost all discussions are equally applicable to both 1.8 GHz and 850 MHz CDMA systems if appropriate changes are made for the frequency-related numerical values (for example, wavelength). Also, the guidelines in this book can be applied to the systems deployed in other parts of the globe, though local regulations and practices may modify some of the specific procedures, as well as the 1.8 GHz or 850 MHz spectral band.

Acknowledgments

We would first like to thank all our customers who reviewed and provided us with invaluable comments on the draft versions of this material. As a compilation of memos, papers, and reports, this book has a long list of contributors. We would like to thank and acknowledge the following Bell Labs authors without whose significant contributions this book would not have been possible: Terry Cheng, Michael Craig, Asif Gandhi, Kelvin Ho, Ching Yao Huang, Raafat Kamel, Terry Lenahan, Allen He, James Lin, Shen-De Lin, Mark Newbury, and Yuqi Yao.

Preface

In addition, major contributions have come from the following people and organizations for the material in the appendices. We have benefited from the members and management of the Wireless Design Center of Lucent Technologies' Bell Labs, who shared with us their insights on and experiences with the RF design process in Appendix A. We thank Neil Bernstein, Victor Dasilva, Jim McElroy, Xiao Wang, and Xiao (Susan) Wu who allowed us to use part of their document in Appendix B and Steven Cosmas, Tzong-Yih Li, Jim Pelech, Shaoming Tong, and Hengxin Weng for their preparation of Appendix C and the demonstration version of the CE4 CD-ROM.

Reviews of our early manuscript by Qi Bi, Zoran Siveski, Hong Yang, and Linda Zeger provided many enhancements reflected in this book. We are grateful to the outside reviewers, Jacques Beneat, Philip H. Enslow, Jr., and Allen H. Levensque, for their constructive and helpful comments. We are indebted to Craig Lawson of Lucent Network Systems Customer Training & Information Products for his fine editorial skills that have improved the readability of this book.

Last, but not least, our special thanks go to Sang Bin Rhee, who inspired us to publish this book, and Hank Menkes, who has provided tremendous support and continuous encouragement throughout this publication project.

Kyoung Il Kim
August 1999, Whippany, N.J.

Foreword

Wireless communications has a rich history, dating back to the fundamental experiments performed by Heinrich Hertz in the 1880s and carried forward by Guglielmo Marconi toward the end of the 19th century. It was Marconi who first demonstrated the practicality of communications with ships at sea—the first instantiation of mobile communications. From then on, the waves of innovation rushed rapidly forward, so that in the United States in 1921, the Detroit police department installed the first mobile radio dispatch system. During this time, Bell Laboratories was conducting research in many aspects of radio transmission. Also, it was in the late 1920s that Harry Nyquist came up with his now-famous sampling theory, a breakthrough key to the digital representation of signals.

Spread spectrum technology has a briefer, but no less fascinating, history. While the initial spread spectrum patents date back to the 1920s, the stimulus for reducing the technology to practice was the need for secure communications to support command and control functions during World War II. The origins of spread spectrum have been documented in a very readable format by Robert Scholtz [1], Robert Price [2] and William Bennett [3].

Bell Laboratories has been in the forefront of innovation in spread spectrum technology and during World War II produced a system for secret telephony designated internally as the X system and code named "Sigsaly" by the United States Signal Corps. Sigsaly used digitized and encoded speech—it was the first application of vocoding and Pulse Code Modulation—spread and scrambled using pseudorandom subcarriers. At the receiver, correlation techniques were used to recover the signals. R. K. Potter and

R. C. Mathes of Bell Laboratories in 1941 filed the earliest patents for the system [4]. Due to the secrecy of the project, the patents were issued only in 1976. During the war, Roosevelt and Churchill used the system for secret discussions. Later, the famed Bell Laboratories researcher, Claude Shannon, showed that the communications channel capacity is maximized by transmission of information carried by a set of noise-like waveforms.

Bell Laboratories continues to be in the forefront of advanced wireless communications. From the invention of the cellular communications concept by D. H. Ring [5] in 1947, the system experiments in the 1960s, the proposal to the U.S. Federal Communications Commission for a wide-area cellular system in 1970, and the cellular trials in the 1970s, Bell Laboratories led the industry in the deployment of the first commercial cellular systems in the 1980s. It should be noted that Bell Laboratories made the first digital cellular call on a live, operational cellular system in 1988. Subsequently Bell Laboratories took a lead role in standardization of CDMA (Code Division Multiple Access) wireless systems, which was completed in 1993.

Today, with a subscriber base of over 350 million worldwide, the industry stands poised on the threshold of third generation wireless technology and services. To carry wireless into the future, Lucent Technologies and Bell Laboratories have amassed a vast portfolio of practical and theoretical expertise in the areas of spread spectrum, CDMA, and wireless mobility systems. Lucent Technologies has deployed more than 160 CDMA networks to over forty service providers worldwide. These systems provide highly reliable, high quality service to millions of users. It is from this experience that we have developed techniques and guidelines for designing, deploying, and optimizing CDMA systems. It is these techniques that, in the tradition of Bell Laboratories, we share with you, the reader.

References

[1]   R. A. Scholtz. The origins of spread-spectrum communications, *IEEE Transactions on Commun.*, vol. COM-30, pp. 822–854, May 1982.

[2]   R. Price. Further notes and anecdotes on spread-spectrum origins, *IEEE Transactions on Commun.*, vol. COM-31, pp. 85–97, Jan. 1983.

[3]   W. R. Bennett. Secret telephony as a historical example of spread-spectrum communication, *IEEE Transactions on Commun.*, vol. COM-31, pp. 98–104, Jan. 1983.

[4]   M. D. Fagen, ed. A history of engineering and science in the Bell System—national service in war and peace, Bell Telephone Laboratories, pp. 296–297, 1978.

[5]   D. H. Ring. Mobile telephony—wide area coverage, Bell Telephone Laboratories Technical Memorandum, Dec. 11, 1947.

George I. Zysman, Ph.D.
Chief Technical Officer
Wireless Networks Group
Bell Labs, Lucent Technologies

CHAPTER 1

Introduction

This book provides basic radio frequency (RF) engineering guidelines and recommendations for use in the planning and design of a code division multiple access (CDMA) personal communication service (PCS) system.

These guidelines are generic. Specific implementations may vary from system to system. All information is consistent with the CDMA PCS common air interface standard (ANSI J-STD-008), minimum performance standard for CDMA PCS base stations (PN3383), cellular CDMA common air interface standard (IS-95A), minimum performance standards for cellular CDMA base stations (IS-97), and minimum performance standards for cellular CDMA mobile stations (IS-98). References 1.1-1.5 list these standards.

This book consists of 11 chapters. Chapter 1 provides introductory material. Chapter 2 presents an overview of CDMA features. It summarizes CDMA concepts and operations, and provides a basis for understanding the sections that follow. Chapter 3 discusses spectrum coordination issues and in particular recommends center frequency assignments for CDMA PCS systems.

ANSI J-STD-008 [1.4] defines a pair of pseudo-noise (PN) sequences used for pilot channels. Each sector of a base station is assigned a specific phase offset (that is, time offset) of these sequences so that the pilot channel of a sector can be distinguished from every other sector by this specific phase offset. Chapter 4 considers the problem of choosing appropriate phase offsets for different sectors. Phase offsets can be reused—recommendations include a lower bound for pilot phase offset increment and a procedure for phase offset assignment.

Chapter 5 addresses the elements of call processing related to mobile station access. In ANSI J-STD-008 [1.4], CDMA mobile stations transmit on the access channels according to a random access protocol. This chapter analyzes the packet throughput and delay performance of the mobile station access protocol in terms of various mobile station access parameters.

A CDMA system supports several types of handoff, including hard handoff, soft handoff, and softer handoff. A mobile station in hard handoff switches from one base station to another base station by a brief interruption of the traffic channel. An example of hard handoff is handoff from one CDMA RF carrier to another CDMA RF carrier. A CDMA-to-CDMA hard handoff can also occur at the boundary between different mobile switching centers of the same CDMA carrier.

Soft handoff is a technique in which a mobile station, while moving between one cell and its neighboring cells, simultaneously transmits and receives the same signal from several base stations. On the forward link, the mobile station in soft handoff can combine the signals using appropriate diversity techniques. On the reverse link, the mobile switching center can decide which base station is receiving the stronger signal.

In softer handoff, neighboring sectors of the same cell support the mobile station's call. Proper use of soft and softer handoff can enhance call quality, improving cell coverage and capacity. In soft handoff, the mobile station continuously scans for pilots and establishes communication with any cell (up to three) whose pilot exceeds a given threshold. Similarly, communication with cells whose pilot drops below a threshold is terminated. Chapter 6 addresses the operation of soft handoff and recommends the values of various soft handoff parameters.

Link budgets show the computation of the ratio of bit energy to thermal noise plus interference power spectral density based on the cell loading, channel activity factor, required transmit power, transmit and receive antenna gain, receiver noise figures, fade margin, and propagation path loss. Chapter 7 presents the methods of developing the link budgets for the forward link and reverse link and provides examples.

A simple definition of capacity is the number of calls by mobile stations that can be supported by the system. Transmit power constraints and the system's self-generated interference ultimately restrict CDMA capacity. The reverse link reaches capacity when a mobile station has insufficient transmit power to overcome the interference from all other mobile stations to meet the required ratio of bit energy to interference power density at the intended base station. Similarly, in the forward link, capacity is reached when the total power required to successfully transmit to all mobile stations hosted by the cell exceeds base station power in order to meet the required ratio of bit energy to interference density at all intended mobile stations. Accordingly, the area in which CDMA coverage can be achieved is comprised of those locations where the bit energy to interference power density ratio requirement can be met. Chapters 8 and 9 discuss the capacity and coverage of CDMA systems.

Prior to CDMA mobile station access, the strongest pilot selected among different CDMA base stations has to be acquired. The pilot channel also aids the handoff operation. Accordingly, the area in which CDMA coverage can be achieved is comprised of those locations where the requirement of the pilot chip energy to noise plus interference power density ratio can also be met. Chapter 9 analyzes the following for both hard and soft handoff: coverage probability, percentage power allocation for the pilot channel, and the required threshold of the pilot chip energy to noise plus interference power spectral density ratio.

In a CDMA system, a channel element (CE) performs the baseband spread spectrum signal processing for a given channel (pilot, sync, paging, or traffic channel). Erlang capacity is determined not only by the maximum number of CEs available for traffic channels, but also by the maximum number of simultaneous active users.

At a three-sector CDMA cell, provisioning CEs per sector based on the per-sector traffic loads would be wasteful because the trunking efficiency that can be gained by pooling the CEs would not be achieved. When all the CEs are pooled, any CE can be assigned, regardless of sector, to any user in the cell. As another factor affecting provisioning, the CDMA system must impose a limit on the number of simultaneous users in a sector to control the interference between users that have the same pilot. Therefore, it would be inappropriate to provision CEs for base stations based just on the total load because such an approach neglects the per-sector limits. The optimum procedure must account for blocking that occurs when users saturate a sector and when all CEs at the base station are busy. To accommodate soft handoff, more CEs need to be allocated. The recommendation is initially for an additional thirty to thirty-five percent, although percentages at individual base stations may need to be adjusted, based on the actual percentage of mobile stations in soft handoff.

Chapter 10 develops an algorithm that calculates the Erlang capacity for a CDMA base station based on the blocking probability requirement, the individual traffic loads of the sectors, and the total number of CEs. Thus, the smallest number of CEs that meets the blocking objective can be determined. To illustrate the use of this algorithm, Chapter 10 presents numerical results of the calculated Erlang capacity per sector and the required number of traffic CEs. These numerical results are based on ten to fifteen users per sector per carrier with a two percent blocking objective.

Chapter 11 presents antenna isolation guidelines for collocated base stations. This chapter examines the antenna separation required between two collocated RF stations based on the criteria of minimizing degradation caused by receiver desensitization and receiver overload and minimizing mutual interference due to intermodulation product.

Appendices A and B discuss some of the practical engineering issues, such as the initial RF design that uses the RF coverage prediction and the RF optimization for fine-tuning of various CDMA system parameters. We have also included a demo version of Cellular Engineering 4 (CE4), a software tool used for RF coverage prediction. Appendix C explains installation and usage of the tool.

One of the important topics for the CDMA system is power control. The fundamental purposes of power control are to maintain voice quality and to increase system capacity. These goals can be achieved by controlling all CDMA signals so they are at the lowest level necessary to meet certain bit energy to interference power density ratios while avoiding unnecessary levels of interference to other signals. On the forward link, power control is implemented with a closed-loop algorithm in which the mobile station requests forward-link power adjustments based on the received frame error rate (FER). On the reverse link, power control is implemented with open- and closed-loop algorithms. In the open-loop path, the mobile station makes power adjustments based on its estimate of path loss from base station to mobile station. In the closed-loop path, the base station compensates for uncorrelated path loss variations and additional sources of interference by estimating the received bit energy to interference power density ratio from the mobile station and transmitting appropriate power adjustment commands. These two control paths jointly determine the final value of mobile station transmit power.

Although the standards specify the details of the operation of the power control, their implementation can differ from one system to another. As such, we do not discuss this important but implementation-specific subject in this book.

1.1 References

[1.1] TIA/EIA/IS-95A. *Mobile Station-Base Station Compatibility Standard for Dual-Mode Wideband Spread Spectrum Cellular System*, March 1995. (To purchase the complete text of any TIA document, call Global Engineering Documents at 1-800-854-7179 (global@his.com) or send a facsimile to 303-397-2740.)

[1.2] TIA/EIA/IS-97. *Recommended Minimum Performance Standards for Base Stations Supporting Dual-Mode Wideband Spread Spectrum Cellular Mobile Stations*, February 1994.

References

[1.3] TIA/EIA/IS-98. *Recommended Minimum Performance Standards for Dual-Mode Wideband Spread Spectrum Cellular Mobile Stations*, February 1994.

[1.4] ANSI J-STD-008. *Mobile Station - Base Station Compatibility Requirements for 1.8 and 2.0 GHz CDMA PCS*, March 1995.

[1.5] PN3383. *Recommended Minimum Performance Requirements for Base Stations Supporting 1.8 to 2.0 GHz CDMA Mobile Stations*, March 1995.

CHAPTER 2

CDMA Overview

2.1 Concept

CDMA is a multiple-access concept based on the use of wideband spread-spectrum techniques that enable the separation of signals that are coincident in time and frequency. All signals share the same spectrum. The energy in each mobile station's signal is spread over the entire bandwidth and coded to appear as broadband noise to every other mobile station. Identification and demodulation of individual signals occur at the receiver by applying a replica of the code used for spreading each signal. This process enhances the signal of interest while dismissing all others as broadband interference. This level of interference rises with the number of mobile stations. Since ensuring call quality requires a minimum signal-to-interference ratio, the total level of background interference ultimately limits system capacity; consequently, careful control of all transmissions ensures that they operate with the least necessary power.

The CDMA concept can be contrasted with other multiple access techniques. In frequency division multiple access (FDMA), each mobile station has full-time use of part of the spectral allocation. The FDMA technique divides the allocation into a number of narrowband portions (channels). Each mobile station confines its signal energy within a channel. Base stations and mobile stations use frequency selective filters to distinguish between signals coincident in time. In time division multiple access (TDMA), each mobile station has part-time use of all the spectral allocation. The TDMA technique breaks down the allocation into a number of time slots. Each mobile station confines its signal energy within a time slot. Base stations and mobile stations use time gating to distinguish signals coincident in frequency. In CDMA, each mobile station has full-time use of the entire spectral allocation and spreads its signal energy over the entire bandwidth. Base stations and mobile stations use codes unique to each signal to distinguish those signals coincident in time and frequency.

2.2 Attributes

The chief rationale behind the deployment of a CDMA system is its potential for high spectral efficiency, that is, its ability to support significantly more mobile stations within a given bandwidth. The design of key system components, such as power control and soft handoff, realizes and enhances this potential while maintaining acceptable call quality. In addition, the modulation concept permits the offering of such desirable system attributes as dynamic capacity and voice privacy.

A summary of the key attributes of a CDMA system follows. More detailed explanations can be found in succeeding chapters.

2.2.1 Capacity

Capacity considerations are fundamental to CDMA planning and operation. For the purpose of this discussion, we will define capacity simply as the number of mobile stations that can be simultaneously supported. We will address forward- and reverse-link capacity separately.

In each link, CDMA signals share the same spectrum (that is, RF carrier). Each mobile station uses a unique code to make its signal appear as broadband interference to every other mobile station. Power control minimizes the impact of this interference by adjusting each signal level to the minimum necessary to achieve desired call quality. In the following sections, application of these principles describes the dynamics of CDMA capacity.

2.2.1.1 *Reverse Link*

To place a call, a CDMA mobile station must have sufficient power to overcome the interference generated by all other CDMA mobile stations within the same band; that is, the received signal at the base station must achieve a required signal-to-interference ratio. The mobile station transmit power required will thus depend on the distance of the mobile station from the base station as well as on the total level of interference (that is, cell loading).

The establishment of each additional call raises the interference levels seen by all mobile stations; accordingly, to maintain call integrity, each mobile station appropriately increments its transmit power. These adjustments, in turn, raise the level of interference that the next mobile station must overcome. This process repeats itself until a new mobile station cannot achieve acceptable voice quality at the base station. The system reaches its capacity at this point.

The capacity limit occurs because the mobile stations eventually have insufficient transmit power to overcome interference levels. Thus, the limit depends on factors that influence the level of interference seen at the base station, for example, traffic distributions inside and outside the cell. Since the mobile station restricts output power when the mobile station user is not speaking, the limit will also depend on the average level of reverse-link voice activity.

The many factors influencing CDMA capacity give rise to a desirable flexibility in system operation. Since capacity depends on interference levels, a cell's capacity is inherently dynamic, that is, a cell can naturally absorb more mobile stations if neighboring cells are lightly loaded. In addition, the system can naturally exploit the reduced levels of interference generated by low voice activity. Finally, capacity limits are soft rather than hard because system capacity can be increased by lowering voice-quality requirements. This procedure supports more mobile stations at the expense of slightly degrading the call quality of all mobile stations.

The flexibility described above makes it difficult to exactly assess the reverse-link capacity that will apply to all situations. As discussed in Section 8.1, a useful reference point can be obtained by assessing the number of allowed mobile stations in an embedded cell when the power control is ideal and the mobile station transmit powers are not limited. The maximum number of mobile stations that can be supported under these circumstances is called the *pole point* or *power pole*.

Section 8.1 tabulates the values used in assessing pole point. For these figures, the pole point is approximately twenty-four. This value applies to a directional cell. Use of 1.23 MHz of the PCS spectral allocation and 13 Kbps vocoder option obtains this pole point. Thus, implementing CDMA within a modest portion of the PCS band results in a CDMA system with significant capacity.

2.2.1.2 Forward Link

Restrictions on base-station radiated power fundamentally determine upper limits on forward-link capacity. The forward-link signal comprises message traffic for subscribers, a sector-specific signal (pilot) used by all mobile stations, and miscellaneous signals (for example, sync, paging). The base station allocates the total power among these functions. Additional mobile stations cannot be supported when the sum of the allocations required exceeds the available transmit power.

The need for a minimum signal-to-interference ratio at each mobile station governs the required allocations. The power allocated to other mobile stations within the cell as well as the received power from neighbor base stations contribute to interference. Use of orthogonal codes partially mitigates this interference because it allows the receiver to suppress signals intended for other mobile stations; however, multipath effects limit the extent to which this interference can be screened out.

The requirement that a generous fraction of power must be allocated to the sector pilot further restricts forward-link power distributions. The sector pilot is important because all mobile stations use it in base-station acquisition and tracking. Capacity limits are, therefore, reached when the remaining power, distributed among all mobile stations, is insufficient to meet mobile station signal-to-interference ratio requirements.

2.2.2 Power Control

As discussed above, capacity can be maximized by minimizing the total level of system interference, that is, by controlling all CDMA signals so they are at the lowest level necessary to meet signal-to-interference ratio requirements. Power control ensures that each signal meets minimum requirements for communication while avoiding undue levels of interference with other signals.

A closed-loop algorithm on the forward link and open- and closed-loop algorithms on the reverse link accomplish control. The measurement of parameters known to influence the desired output forms the basis of the open-loop mechanisms whereas the direct measurement of the output itself forms the basis of the closed-loop mechanisms.

Reverse-link control ensures the minimum necessary signal-to-interference ratio at the base station for each mobile station. In the open-loop path, the mobile station makes power adjustments based on its estimate of path-loss from base station to mobile station. The mobile station's measurement of received total power forms the basis for this estimate. These adjustments compensate for path-loss variations that are correlated between the forward and reverse links. In the closed-loop path, the base station compensates for uncorrelated path-loss variations (for example, multipath fading) and additional sources of interference by estimating the received signal-to-interference ratio from the mobile station and transmitting appropriate power adjustment commands. The final value of mobile station transmit power is jointly determined by these two control paths.

Forward-link control ensures the minimum necessary signal-to-interference ratio at each mobile station. In this closed-loop mechanism, the mobile station requests forward-link power adjustments based on its received frame error rate.

2.2.3 Soft Handoff

CDMA provides various mechanisms to ensure a robust handoff, that is, to ensure call support when a mobile station crosses the boundary from one cell to another. The chief mechanism employed is *soft* handoff in which the mobile station's call is simultaneously supported by up to three sectors. This process enables the mobile station to establish contact with the sectors it is likely to travel through well before it leaves its serving (host) base station. In addition, the simultaneous support provides a diversity gain that improves link quality in *fringe* areas. The application of power control from neighbor base stations also ensures that a progressively distant mobile station will not unduly boost its transmit strength and become a primary source of interference to a nearby base station.

The CDMA soft handoff differs from the more familiar *hard* handoff in several ways. Hard handoff requires a brief interruption of the traffic channel. CDMA soft handoff does not require channel switching per se because every cell reuses the same channel (carrier). Moreover, the acquisition of new base stations takes place before contact with the old (serving) base station breaks off. No interruption of the traffic channel occurs. This handoff procedure is robust because the mobile station connects with the new host(s) before disconnecting from the old host(s). This process is often referred to as a make-before-break connection as opposed to the break-before-make.

2.2.4 Voice Activity

Exploitation of voice activity may enhance capacity. On average, each link in a two-way voice conversation is active about half the time. If transmitters vary output power with voice activity, the total interference power from a large number of mobile stations will be reduced by about a factor of two.[1] This reduction in interference translates naturally into a direct increase in system capacity. This use of voice activity is possible because all subscribers reuse the same channel. In contrast, an analog frequency modulation (FM) system could exploit voice activity only through the difficult task of reassigning the channel resource whenever the speaker pauses.

---

1. On the forward link, the actual reduction depends on the level of *channel* activity, which includes both voice activity and message activity (for example, the energy level of punctured power control bits—see section 3.1.3.1.8 of [2.1]).

2.3 Reference

[2.1] ANSI J-STD-008. *Mobile Station - Base Station Compatibility Requirements for 1.8 and 2.0 GHz CDMA PCS*, March 1995.

Chapter 3

Spectrum Coordination

3.1 Spectrum Coordination

Table 3–1 defines the 120 MHz of PCS spectrum. It comprises two separate 60 MHz bands, 1850-1910 MHz for mobile station (uplink) transmission and 1930-1990 MHz for base station (downlink) transmission. These two bands are each divided up into six frequency blocks, A through F. A through C are 15 MHz (15 MHz uplink and 15 MHz downlink) blocks and D through F are 5 MHz blocks. The table lists the blocks in order of increasing frequency. Thus, frequency block D has two 5 MHz blocks contiguous with the 15 MHz (uplink and downlink) blocks of frequency allocated to blocks A and B.

Table 3–1 PCS Spectrum Allocation

Block Designator	Bandwidth Allocated (MHz)	Transmit Frequency Band (MHz)	
		Mobile Station	Base Station
A (MTA)	30 (15/15)	1850 – 1865	1930 – 1945
D (BTA)	10 (5/5)	1865 – 1870	1945 – 1950
B (MTA)	30 (15/15)	1870 – 1885	1950 – 1965
E (BTA)	10 (5/5)	1885 – 1890	1965 – 1970
F (BTA)	10 (5/5)	1890 – 1895	1970 – 1975
C (BTA)	30 (15/15)	1895 – 1910	1975 – 1990

Table 3–2 CDMA Channel Numbers and Corresponding Frequencies for Band Class 1

	CDMA PCS Channel Number (N)	CDMA PCS Channel Frequency (MHz)	TDMA PCS Channel Number (N)	TDMA PCS Channel Frequency (MHz)
Base Station Receive	N = 0 to 1199	$1850.000 + 0.050 \times N$	N = 1 to 1999	$1849.980 + 0.030 \times N$
Base Station Transmit	N = 0 to 1199	$1930.000 + 0.050 \times N$	N = 1 to 1999	$1930.020 + 0.030 \times N$

Reproduced from [3.1] under written permission of the copyright holder (Telecommunications Industry Association)

The 120 MHz (1850-1910 and 1930-1990) of spectrum shown in Table 3–1 has been allocated for licensed terrestrial systems; the 20 MHz (1910-1930) of spectrum in between has been allocated for unlicensed applications. Note that frequency blocks A and B are designated for use in the fifty-one major trading areas (MTAs), while blocks C through F are for the 492 basic trading areas (BTAs). The frequency allocations and block designations shown in Table 3–1 are common to both the CDMA and TDMA technologies used for PCS. The subsections below indicate the differences between CDMA and TDMA uses of the spectrum.

3.1.1 Spectrum Allocation and Channel Numbering

The channel numbering scheme for CDMA PCS provides for a straightforward translation between channel number and carrier frequency. Table 3–2 shows the relationship for both base station transmit and receive frequencies. The 50 KHz channel spacing over the two 60 MHz bands for transmit and receive results in a total of 1200 CDMA channel numbers.

Table 3–2 also shows the translation of TDMA channel number to frequency. The TDMA values emphasize the fact that similar channel numbers for CDMA and TDMA systems do not translate into the same carrier frequencies. This is important to personnel concerned with frequency planning when they try to coordinate and/or control interference for a CDMA PCS system that shares a geographic border with a TDMA system utilizing the same frequency block.

Note that while the *Base Station Transmit* and *Base Station Receive* frequencies for a given channel number will differ by 80 MHz in CDMA, the transmit and receive frequencies will differ by 80.04 MHz in TDMA. Thus, questions of interference from and to TDMA channels will have to be considered for both the uplink and downlink because the calculation of TDMA channel numbers, which should not present interference problems to or from the CDMA system, will differ depending on whether the derivation of channel number is for uplink or downlink frequencies. Obviously, the only usable frequencies are those that are free from interference on both uplink and downlink.

3.1.2 Channel Availability

Although the CDMA channel-numbering algorithm with 50 KHz channel spacing implies the availability of 1200 CDMA carriers, not all 1200 are actually usable. Table 3–3 indicates the availability of the channels by classifying them as valid (usable) channels, conditionally valid, or not valid.

The designation of channels 0 to 24 and 1176 to 1199 as not valid eliminates the possibility of interference between PCS systems and the services allocated to the spectrum above, below, and between the two 60 MHz spectrum allocations comprising the PCS spectrum.

The channels specified as conditionally valid are the twenty-five lowest (except for Block A) and the twenty-five highest (except for Block C) channels in each block. These channels are valid only if the service provider also owns the adjacent block of the spectrum.

Looking at it another way, all channels are valid for use as CDMA carriers except for the twenty-five lowest channels and the twenty-five highest channels in each block. Thus, 251 channels are unconditionally available (that is, "valid") for designation as carrier frequencies in frequency blocks A, B, and C, and fifty-one are unconditionally available channels for Blocks D, E, and F. If a service provider were to obtain licenses in two adjacent blocks, then an additional fifty channels would become available from the conditionally available channels.

Not all of the valid and conditionally valid channels can be used simultaneously as carriers in a given system. Once a channel number has been specified for use as the first carrier in a system, there are minimum spacing rules for carriers in use that limit how close the new carrier can be above or below the previously existing carrier(s). While the classification of channels as valid and conditionally valid is by Federal Communication Commission (FCC) decree, CDMA technology considerations determine the minimum spacing between active carriers. Generally, the minimum carrier spacing of twenty-five CDMA channels dictates channel specification, which is consistent with the nominal 1.25 MHz bandwidth for CDMA.

Table 3-3 CDMA Channel Allocation Availability

Frequency Block	CDMA Frequency Assignment Validity	CDMA Channel Number	Transmit Frequency (MHz) Personal Station	Transmit Frequency (MHz) Base Station
A (15 MHz)	Not Valid	0–24	1850.000–1851.200	1930.000–1931.200
A (15 MHz)	Valid	25–275	1851.250–1863.750	1931.250–1943.750
A (15 MHz)	Conditionally Valid	276–299	1863.800–1864.950	1943.800–1944.950
D (5 MHz)	Conditionally Valid	300–324	1865.000–1866.200	1945.000–1946.200
D (5 MHz)	Valid	325–375	1866.250–1868.750	1945.600–1948.750
D (5 MHz)	Conditionally Valid	376–399	1868.800–1869.950	1948.800–1949.950
B (15 MHz)	Conditionally Valid	400–424	1870.000–1871.200	1950.000–1951.200
B (15 MHz)	Valid	425–675	1871.250–1883.750	1951.250–1963.750
B (15 MHz)	Conditionally Valid	676–699	1883.800–1884.950	1963.800–1964.950
E (5 MHz)	Conditionally Valid	700–724	1885.000–1886.200	1965.000–1966.200
E (5 MHz)	Valid	725–775	1886.250–1888.750	1966.250–1968.750
E (5 MHz)	Conditionally Valid	776–799	1888.800–1889.950	1968.800–1969.950
F (5 MHz)	Conditionally Valid	800–824	1890.000–1891.200	1970.000–1971.200
F (5 MHz)	Valid	825–875	1891.250–1893.750	1971.250–1973.750
F (5 MHz)	Conditionally Valid	876–899	1893.800–1894.950	1973.800–1974.950
C (15 MHz)	Conditionally Valid	900–924	1895.000–1896.200	1975.000–1976.200
C (15 MHz)	Valid	925–1175	1896.250–1908.750	1976.250–1988.750
C (15 MHz)	Not Valid	1176–1199	1908.800–1909.950	1988.800–1989.950

3.1.3 Preferred Channels

The preceding subsection specified the channels that are valid, or at least conditionally valid, carrier frequencies that the service provider can specify for use in the system's frequency plan. Issues dealing with inter-system or intra-system interference might dictate the selection of these frequencies. If these issues are insignificant factors in the system performance, the number of channels that the service provider might con-

Table 3–4 Preferred CDMA Channels

Frequency Block	Preferred Channel Numbers
A	25, 50, 75, 100, 125, 150, 175, 200, 225, 250, 275
D	325, 350, 375
B	425, 450, 475, 500, 525, 550, 575, 600, 625, 650, 675
E	725, 750, 775
F	825, 850, 875
C	925, 950, 975, 1000, 1025, 1050, 1075, 1100, 1125, 1150, 1175

sider for carrier frequencies can be reduced significantly to the list of *preferred channels* in Table 3–4. These are the channel numbers that a mobile station will *scan* when looking for service. Thus, a system must use at least one (or more) of these carriers at each site in the system if the sites are to be capable of providing (CDMA) service.

Conditionally valid channels 300, 400, 700, 800, and 900 can only be used if the service provider has licenses for both the frequency block containing the channel and the immediately adjacent frequency block (for example, channel 300 is a likely carrier channel if the service provider has licenses for both blocks A and D). This accounts for their exclusion from the above list. Conditionally valid channels should be used for traffic only and not access.

3.1.4 Intra-System Frequency Planning Issues

In discussing CDMA frequency planning issues, description of assignments is in per-base-station terms. Note that, for directional base stations, the rules and/or statements are equally valid for the individual base station faces.

Because of $N = 1$ frequency reuse, CDMA frequency planning is highly simplified compared to analog systems. For systems requiring only one carrier per base station, the carrier channel must be one of the preferred channels listed in Section 3.1.3. Under normal circumstances, there is no reason to consider any one of the preferred channels (associated with the service provider's frequency block) as having any specific advantage; thus, the first channel assignment for a new system is a *random choice* from the list of preferred channels. The same channel should be used in all the base stations throughout the system to take advantage of soft and softer handoff capabilities.

Local interference (in both geographic and frequency spectrum terms) is one possible exception to the above assumption of there being no difference between the preferred channels. This interference might be caused by (or in) one of the services that used part of the PCS spectrum prior to the PCS allocation. Section 3.1.5.3 discusses channel assignments in this environment.

As the number of CDMA subscribers increases, growth can occur through the addition of new base stations or the increase of CDMA carriers per base station. Assuming that additional carriers can still be assigned and that there is no reason (such as the possibility of improved coverage performance) to add new base stations, then the logical means of system growth is through the addition of CDMA carriers per base station. The additional CDMA channel assignments do not have to be taken from the list of preferred channels. However, except for outside interference, there is really no reason to assign channels other than the preferred channels. Assigning preferred channels makes optimum use of the available spectrum by the closest spacing of carriers as dictated by the CDMA signal bandwidth.

Because of non-uniform subscriber growth throughout the system, it is likely that not every base station in the system will have the same number of carriers. In these instances, the base stations having the extra carrier will not be able to take advantage of soft and/or softer handoff at those boundaries where use of the carrier is on one side only. Except in extreme differences in subscriber growth, the number of carriers in use as one crosses coverage boundaries should not differ by more than one carrier. Except for this additional carrier, the channel assignments on both sides of a base station coverage boundary should be the same in order to take maximum advantage of soft and softer handoff.

3.1.5 Inter-System Frequency Planning Issues

The sections that follow address inter-system frequency planning issues as a two-system problem, that is, they discuss frequency planning issues/concerns for a CDMA system that is operating in the same area or in an area adjacent to that of a second system. The second system may or may not be a CDMA system; the assumption, however, is that the second system uses at least part of the same frequency block as the CDMA system of concern. Another assumption is that the rules for the use of the conditionally valid channels (for example, channels 276 to 299 in the A block and channels 300 to 324 in the B block can only be used if the service provider is licensed for both frequency blocks) eliminate problems of interference between channel assignments in the frequency block of concern and any (CDMA or TDMA) channel assignments for systems using the adjacent frequency blocks. It is also assumed that cases with boundaries involving three or more systems are rare and, should they occur, could be treated by simple extension of the cases considered in the following subsections.

Spectrum Coordination

3.1.5.1 *CDMA in Both Systems*

Because CDMA uses $N = 1$ frequency reuse, channel selection should not be affected for base stations on the boundary if the geographically neighboring system is also a CDMA system. However, pilot phase offset assignments, as discussed in Chapter 4, would have to be coordinated.

3.1.5.2 *CDMA in One System, TDMA in the Second*

If the second system is a TDMA system operating in an adjacent geographic area and using the same frequency block, coordination of frequency utilization will be required for base stations along the boundary. This coordination will also include the establishment of a guard zone.[1]

Although coordination of frequency utilization could result in any division of the spectrum to which both service providers agree, the most obvious one that will be assumed here would provide equal spectrum allocations to the affected base stations on both sides of the system boundaries. Thus, frequency block A, B, or C service providers would each have spectrum allocations of 7.5 MHz (per uplink and per downlink) for use on their respective sides of the system boundary; similarly, frequency block D, E, and F service providers would each have 2.5 MHz of spectrum for use at their system boundary sites.

Because CDMA systems require that at least one of the channel assignments at each site be a preferred channel, it is impossible to evenly divide up a frequency block by giving one service provider the top half of a frequency block and the remaining bottom half to the second service provider. As the listing of preferred channels in Section 3.1.3 demonstrates, frequency block A, B, and C service providers each have a total of eleven preferred channels while the frequency block D, E, and F service providers have a total of three each. To provide approximately equal spectrum allocations for both CDMA and TDMA along the system boundaries, frequency block A, B, and C CDMA service providers should take the portion of the spectrum corresponding to any six *consecutive* (for example, 25, 50, 75, 100, 125, and 150 in block A) preferred channels. For this block A example, the portion of the spectrum available to the TDMA system would then be TDMA channels 1 to 11 and 281 to 498 (or to be accurate, TDMA channels 2 to 12 and 281 to 498 because TDMA channel 1 is not a valid channel).

---

1. For the discussion of guard zone between analog AMPS and CDMA cellular systems operating in 850 MHz band, see, for example, Reference [3.2].

In determining the interference-free TDMA channels above, it was assumed that the CDMA bandwidth was 1.23 MHz and that a 0.27 MHz additional guard band was required. For CDMA preferred channel 25, the carrier frequencies are

- 1851.25 MHz downlink,
- 1931.25 MHz uplink,

minus half the CDMA bandwidth (that is, 0.615 MHz) and the 0.27 MHz results in a CDMA boundary frequency of

- 1850.365 MHz downlink,
- 1930.365 MHz uplink.

If non-integer channel numbers were allowed, the boundary frequencies above would correspond to TDMA channels 12.833 on the downlink and 11.5 on the uplink. Assuming a required half channel bandwidth (taken away from the last two TDMA channel numbers) and then truncating to integer values results in interference-free (for both CDMA and TDMA) TDMA channels in the band

- 1 to 12 for downlink,
- 1 to 11 for uplink.

This reduces to TDMA channels 1 to 11 because channel 12 is interference free only for the downlink.

For frequency block D, E, and F service providers, the CDMA service provider would take the portion of the spectrum associated with any two *consecutive* preferred channels; assignment of TDMA channels would then be from the remaining portions of the frequency block above and below the CDMA allocation. In dividing the spectrum, as assumed above, remember that the CDMA and TDMA channel-numbering algorithms are different and care must be taken when trying to associate the corresponding channel numbers and frequencies of concern between the two systems.

Assuming that the guard zone consists of one tier of cells, then only the base stations immediately on each side of the border between the two systems will have to operate with a reduced spectrum allocation. All cells at least one tier away from the border of concern can have channel assignments made with no concern for inter-system interference.

In the previous discussion, the alert reader will have noticed that the proposed method does not really result in a 50/50 split of the spectrum due to the effects of the 0.27 MHz guard band on each side of the assumed CDMA block of frequencies. Thus, the TDMA service provider does not wind up with half of the 497 usable TDMA channels (that is, half of channels 2 through 498); instead of having at least 248 channels, the TDMA service provider has only 228 channels.

As a possible solution to this problem, the CDMA service provider could use one or two non-preferred carrier frequencies on the border. Thus, instead of using CDMA channels 25 and 150 as the lowest and highest carriers as assumed in the example above, the CDMA service provider might have restored the TDMA's rightful share of half the available frequency block by using carriers 29 and 146. Therefore, the Erlang capacity of the base station faces in question would be reduced for channels 29 and 146 and the adjacent carriers, 50 and 125, on the same faces.

3.1.5.3 Non-PCS Interference

Local interference (in both geographical and frequency spectrum terms) is one possible exception to the statement in Section 3.1.4 that there is no difference between the preferred channels. This interference might be caused by (or in) one of the services that used part of the PCS spectrum prior to the PCS allocation. Under these circumstances, the channel of choice would be one that avoided the spectrum area of concern and could be used throughout the geographic service area. Usability throughout the geographic service area is not a requirement but is desirable because it allows the use of soft and softer handoff.

It is difficult to give specific rules that will apply in all cases. Each case must be considered individually for frequency planning in the event of non-PCS interference. The only hard and fast rule in these cases is that at least one of the channels selected as a carrier must be a preferred channel. Any base station not having at least one preferred channel cannot be accessed (except by handoff from another base station).

The severity of non-PCS interference will depend on whether the non-PCS service is in the mobile station band of transmit frequencies or the base station frequencies; it will also depend on the relative locations of the CDMA base station and the transmit/receive path of the non-PCS service. For instance, the fact that the non-PCS service might consist of a highly directional microwave link might allow the two services to share a common spectrum because the interference from a PCS base station might be dropped to a greatly reduced antenna side-lobe if the base station is positioned properly. In this case, there can still be concern for possible interference received by the mobile stations. However, if the non-PCS interference is localized enough due to a short transmit/receive path and/or highly directional non-PCS antennas, then the interference might still be acceptable in terms of service in the overall coverage area.

If the interference is severe, then the only way to eliminate the problem is through the frequency plan. Thus, only CDMA carriers that prevent mutual interference between the two systems can be used. In determining whether a CDMA channel can be used in an area where there are concerns with non-PCS interference, the CDMA carrier and the non-PCS carrier must be separated sufficiently to at least prevent overlap of their nominal signal bands. If either of the PCS and non-PCS signals is strong enough, then an additional guard band may be required and the two carriers separated even further. It is in such situations that the frequency planner might (and might have to) opt for specifying carrier channels other than those from the preferred channel list. Depending on the carrier frequency and bandwidth of the non-PCS service, the frequency planner must select at least one CDMA carrier from the preferred channel list that will provide interference-free access to the base station. The requirement of more carriers to provide capacity after the selection of at least this minimal number of preferred channels allows the frequency planner freedom to select the combination of CDMA carriers (from the remaining preferred and non-preferred carriers) that will best satisfy goals to maximize the number of usable carriers and/or minimize the potential for interference.

3.2 References

[3.1]   ANSI J-STD-008. *Mobile Station - Base Station Compatibility Requirements for 1.8 and 2.0 GHz CDMA PCS*, March 1995.

[3.2]   K. I. Kim. CDMA Cellular Engineering Issues, *IEEE Trans. Vehicular Technology*, vol. 42, pp. 345-350, August 1993.

CHAPTER 4

Pilot Assignment

4.1 Introduction

IS-95A (Reference [4.1]) and ANSI J-STD-008 ([Reference [4.2]) define a pair of binary sequences, called the I and Q pilot PN sequences, that the base station and mobile station respectively use as spreading-sequences for forward- and reverse-link transmissions in the spread spectrum cellular system. Each sector of a base station is assigned a specific time (or phase) offset of these sequences; this specific time offset distinguishes the transmissions from different sectors. This chapter considers the problem of choosing appropriate time offsets for different sectors in a CDMA system. Due to its similarity to the frequency assignment problem encountered in analog cellular systems, this problem is called the phase assignment problem for CDMA cellular systems.

This chapter is organized as follows. In Section 4.2, we describe a simple model for the pilot signal transmitted by a base station sector. Section 4.3 summarizes several processes of the mobile station that may be impacted by the phase offsets assigned to different sectors in the system. In Section 4.4, we derive the recommended minimum separation between two different pilot PN phase offsets. This section also includes a discussion of the reuse of pilot PN phase offsets. Finally, Section 4.5 outlines a procedure for phase assignment.

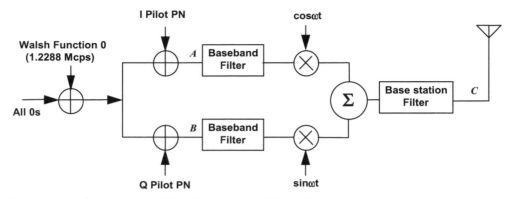

Figure 4–1 Generating a Pilot Signal for a CDMA Cellular System

4.2 Pilot Channel Model

On the forward link, a CDMA carrier consists of the following code channels: the pilot channel, up to one sync channel, up to seven paging channels, and forward traffic channels. Each of these code channels is orthogonally spread by the appropriate Walsh function and then spread by the quadrature pair of PN sequences (the I and Q pilot PN sequences) at a fixed chip rate of 1.2288 Mcps (million chips/sec).

A mobile station operating within the coverage area of the sector uses the pilot channel for synchronization. The sector on each active CDMA carrier frequency transmits the pilot channel at all times. Figure 4–1 is a functional representation of the generation of a pilot signal.

The pilot signal uses the Walsh function **0** consisting of all 0s; hence, the pilot signal is an unmodulated spread spectrum signal that is spread by a quadrature pair of PN sequences. The baseband filters shown in the figure represent both the baseband filtering and the phase equalization as specified in Reference [4.1] and Reference [4.2]. The base station filter shown in the figure represents any amplification and additional filtering carried out at the base station.

Let $\{c_I^{(0)}\}$ and $\{c_Q^{(0)}\}$ denote, respectively, the I and Q pilot PN sequences. The superscript represents the phase shift of the I (or Q) pilot sequence. The elements of the (zero offset) I pilot PN sequence are given by

$$\left(\ldots, c_I^{(0)}(0), c_I^{(0)}(1), c_I^{(0)}(2), \ldots\right), \ c_I^{(0)}(j) \in \{0,1\}, \tag{4.1}$$

and the elements of the (zero offset) Q pilot PN sequence are similarly given by

$$\left(\ldots, c_Q^{(0)}(0), c_Q^{(0)}(1), c_Q^{(0)}(2), \ldots\right), \ c_Q^{(0)}(j) \in \{0,1\}. \tag{4.2}$$

Pilot Channel Model

Let $\{c_I^{(k)}\}$ denote the kth phase offset (equivalently the kth right shift) of the sequence $\{c_I^{(0)}\}$, that is,

$$c_I^{(k)}(i) = c_I^{(0)}(i-k), \quad i = \ldots, -1, 0, 1, \ldots \quad (4.3)$$

The pilot PN sequences are maximal length linear feedback shift register sequences of period $(2^{15} - 1)$ with an additional 0 inserted after a run of fourteen 0s in each sequence. Thus, each sequence has a period equal to 2^{15}. The characteristic polynomials used to generate the sequences and other details may be found in [4.1] and [4.2].

Application of the following assumptions obtains a pilot signal model:

- perfect synchronization of the carrier frequency and phase at the mobile station receiver is assumed, which justifies considering a baseband model for the system, and
- the baseband response of the transmit path to a digital impulse is assumed to be $\Pi(t)$, a rectangular pulse of width T_c.

Let $c_I^{(0)}(t)$ denote the normalized (unit power) and idealized baseband signal corresponding to the I pilot signal with zero phase offset. Thus, $c_I^{(0)}(t)$ is a pulse train given by

$$c_I^{(0)}(t) = \sum_{j=-\infty}^{j=\infty} (-1)^{c_I^{(0)}(j)} \Pi(t - jT_c),$$

where

$$\Pi(t - jT_c) = \begin{cases} 1, & jT_c < t \leq (j+1)T_c \\ 0, & elsewhere. \end{cases} \quad (4.4)$$

Thus, the normalized (unit power) and idealized baseband I pilot signal corresponding to a sector with pilot PN sequence phase offset k is given by the following equation (see Figure 4–2)

$$c_I^{(k)}(t) = \sum_{j=-\infty}^{j=\infty} (-1)^{c_I^{(k)}(j)} \Pi(t - jT_c) = c_I^{(0)}(t - kT_c). \quad (4.5)$$

The most important aspects of the cellular system for phase assignment are the relative powers and time shifts of the pilot signals of the different sectors as measured at the input to a mobile station. The power and time shift of the I and Q pilot signals (for a specific multipath) are equal. Therefore, for phase assignment, it suffices to consider

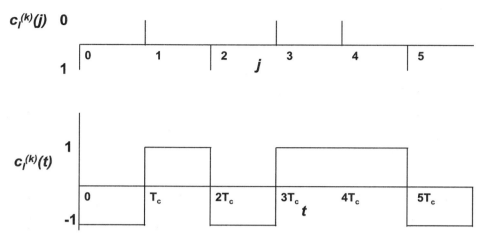

Figure 4-2 Generating $c_I^{(k)}(t)$ from $c_I^{(k)}(j)$

any one of the quadrature (I or Q) pilot signals received at the mobile station. In the rest of this section, it is assumed that the pilot signal transmitted by a sector with pilot PN sequence offset k and power P is given by $\sqrt{P}c_I^{(k)}(t)$. Thus, if τ and L denote the delay in seconds and the effective path loss, respectively, of a specific multipath of the CDMA carrier corresponding to the above pilot, the pilot signal measured at the input of the mobile station (for the specific multipath) is given by

$$\sqrt{\frac{P}{L}}c_I^{(k)}(t-\tau) = \sqrt{\frac{P}{L}}c_I^{(0)}(t-kT_c-\tau). \tag{4.6}$$

4.3 Call Processing for a Mobile Station

This section summarizes several mobile station processes that may be impacted by the phase offsets assigned to different sectors in the system.

1. *Mobile station initialization state:* When a mobile station is powered up, it selects a system to use. If the selected system is a CDMA system, the mobile station proceeds to acquire the system pilot signal and then synchronize to the CDMA system. Usually, the process of acquiring the pilot signal consists of a serial search by the mobile station through the possible time offsets of the pilot signal until it finds a pilot signal at a particular offset. Because the pilot signal is periodic with period $2^{15}T_c$, this time offset can be any value in the range $(0, 2^{15}T_c]$. Typically, the mobile station steps through this range in discrete steps

of T_c sec until it finds the correct time offset (to within $T_c/2$ sec). The mobile station can communicate with the sector only after it has acquired the base station's pilot signal.

If there is only one pilot signal strong enough for the mobile station to detect (that is, above a certain pre-fixed threshold), it follows that the mobile station, on average, would search through half of the total time offset ($2^{14}T_c$) before it reaches the *correct* offset. However, in a typical cellular system, a significant portion of the total area has more than one pilot signal that is strong enough for the mobile station to acquire (for example, near the boundary of two cells). In that case, the average time that a mobile station takes before reaching the offset of any one of the pilot signals depends on the relative phase shifts of the pilot signals. Thus, a judicious choice of the phase offsets of nearby sectors reduces the length of the pilot channel acquisition substate.

2. *Mobile station control on the traffic channel state*: In this state, the mobile station communicates with the sector using the forward and reverse traffic channels. During this state, the choice of phase offsets at different base stations affects several processes at the mobile station. Some of these are listed below.

   - *Searching for other pilot signals*: The mobile station maintains four sets of pilots while it is communicating with the sector: the active set, the candidate set, the neighbor set, and the remaining set. The active set consists of the pilots that the mobile station is currently using for demodulation. The pilots in the other three sets are possible handoff candidates. (For more details, see Chapter 6 and References [4.1] and [4.2].) The mobile station constantly scans these pilots to see if any of them is strong enough to be a candidate for handoff. While their phase offsets individually specify the pilots in the candidate and neighbor sets, a system parameter called *PILOT_INC*, which can be assigned any non-negative integer value less than sixteen, specifies those in the remaining set. The definition of the pilots in the remaining set is all the pilots with offsets that are multiples of $64 \times$ *PILOT_INC* that are not in any of the other three sets. Thus, a high value of *PILOT_INC* leads to a small number of pilots in the remaining set, which leads to a reduction in the time that the mobile station takes to scan for pilots. This therefore decreases the probability of missing a strong pilot in a dynamic environment. However, this improvement is expected to be marginal because pilots in the remaining set are the *lowest priority* pilots in the scan for strong pilots.
   - *Pilot PN phase measurements*: When the mobile station does find pilot signals stronger than a pre-specified threshold (*T_ADD*), it reports the mea-

sured strength and arrival phase of these pilots to the serving sector via the pilot strength measurement message (PSMM). The arrival phase of each pilot signal is computed as described in References [4.1] and [4.2]. The serving sector uses this reported phase to identify the sector that is transmitting the pilot and initiates handoff procedures if necessary. Hence, the pilot PN phases for sectors that are possible candidates for handoff should be assigned so that the serving sector can uniquely identify their transmissions from a measurement of the received phase by the mobile station.

Typically, when demodulating the forward link signal, the mobile station uses coherent combining of the strongest three multipaths of the pilots in its current active set (that is, a Rake receiver). The mobile station finds a strong multipath by checking the strength of the pilot signal corresponding to the multipath. Note that the pilot signals of all base stations in a cellular system are just time shifted versions of each other, and therefore, with the appropriate time delay, an *incorrect* pilot can be mistaken for a pilot in the active set and used for demodulation. However, with the appropriate assignment of pilot PN phases, the probability of this event occurring may be controlled. This is potentially the most harmful effect that may result from a poorly planned pilot PN phase assignment. The next sections analyze this situation in more detail.

4.4 *PILOT_INC* and Pilot Phase Offset Reuse

To avoid interfering pilots from other base stations, the system designer has to properly specify the parameter *PILOT_INC*, the amount of time shift between the pilot signals of different base stations. We can approach the problem of finding an optimum value of *PILOT_INC* by considering two different issues. The first issue is to consider the minimum separation needed between two different pilots, from which a lower limit of *PILOT_INC* can be derived. The second is to look at the reuse distance between the two base stations using pilot signals with the same offset index. From the second consideration, an upper limit can be found. A summary of the criteria that will be used for these appears below.

The separation required for two different pilot signals is based on the following criteria.

S1. The pilot from other sectors that have different PN offsets must not show up in the active search window even with a very large differential delay.
S2. There must be no confusion of pilots in the neighbor search window. That is, in the neighbor search window for neighbor A, signals from other sectors

(with different PN offset, for example, neighbor B) must not show up within the window even with a very large differential delay.

The reuse distance between two base stations having the same PN offset index is based on the following criteria.

R1. The interference from the distant sector that uses the same PN offset must be lower than a certain threshold.
R2. Delay from the distant sector using the same PN offset compared to the delay from the serving sector should be greater than half the size of the active search window such that the undesired finger is not seen inside the active search window.
R3. Sectors with the same PN offset must not show up together in the neighbor search window (for that offset) that is used by any other sector.

Note that criterion R2 is not required if criterion R1 is met because the presence of the undesired finger in the active search window will be ignored for its weak E_c/I_o.[1] However, criterion R2 is included above to guarantee the absence of the undesired finger.

In the following sections, we will consider the above criteria one by one and will derive from each a bound for the pilot PN offset difference.

4.4.1 Pilot Separation and a Lower Limit for *PILOT_INC*

Consider the situation in which a pilot signal that is not in the active set of pilots (or that belongs to a different sector) is mistaken for a pilot belonging to the active set and, therefore, can potentially be used to demodulate an incorrect forward traffic channel. Because all the pilot signals in a system are time-shifted versions of each other, with appropriate time delay, a pilot from any sector can appear to belong to any other sector. However, a greater time delay between a sector and a mobile station implies a greater path loss between the sector and the mobile station and, therefore, a weaker pilot signal at the mobile station input. Thus, if the pilot PN sequence offsets of different sectors have a large separation between them, a pilot signal would have a very high path loss before it appears within the active search window of another pilot.

Specifically, consider the situation in Figure 4–3, in which the pilot from cell 1 appears to be a pilot from cell 2 to a mobile station in cell 2. We will use the following notations throughout this section:

---

1. See (6.1) for the definition of pilot E_c/I_o.

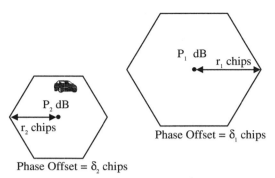

Figure 4–3 Interfering Pilots from Different Base Station

- r_i denotes the radius of the base station i,
- δ_i denotes the pilot PN phase offset associated with the base station i,
- τ_i denotes the time delay from the base station i to the mobile,
- s_i^A denotes the (one-sided[2]) active search window size of the base station i,
- s^N denotes the (one-sided) neighbor search window sizes.

All of the above quantities are expressed in chips.[3]

Thus, $\theta_1(t)$ and $\theta_2(t)$, the pilot signals transmitted by cell 1 and cell 2, respectively, are given by

$$s_1(t) = \sqrt{P_1}\, c_I^{(\delta_1 T_c)}(t) = \sqrt{P_1}\, c_I^{(0)}(t - \delta_1 T_c),$$
$$s_2(t) = \sqrt{P_2}\, c_I^{(\delta_2 T_c)}(t) = \sqrt{P_2}\, c_I^{(0)}(t - \delta_2 T_c). \qquad (4.7)$$

Let L_1 and $\tau_1 T_c$ denote the path loss and time delay, respectively, from cell 1 to the mobile station (of a specific multipath of the pilot signal of cell 1). Similarly, let L_2 and $\tau_2 T_c$ denote the path loss and time delay, respectively, from cell 2 to the mobile station (of a specific multipath of the pilot signal of cell 2).

Thus, $y_1(t)$ and $y_2(t)$, the two pilots received at the mobile station, are given by

---

2. Here, one-sided means a half, that is, s_i^A equals a half of the *SRCH_WIN_A* in number of chips.
3. To derive the conversion factor between the distance in miles to the distance in chips, let d miles denote the distance between two points and let κ chips denote the same distance in chip lengths. Note that

$$1 \text{ chip} \Leftrightarrow 10^{-6}/1.2288 \text{ sec}$$
$$1 \text{ mile} \Leftrightarrow 5.36 \times 10^{-6} \text{ sec} \Leftrightarrow 6.6 \text{ chips}$$
$$\text{or } \kappa = 6.6 \times d$$

PILOT_INC and Pilot Phase Offset Reuse

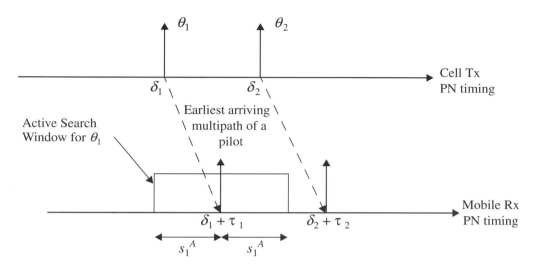

Figure 4–4 Timing Diagram for Criterion S1

$$y_1(t) = \frac{s_1(t - \tau_1 T_c)}{\sqrt{L_1}} = \sqrt{\frac{P_1}{L_1}} c_I^{(0)}(t - \delta_1 T_c - \tau_1 T_c)$$

$$y_2(t) = \frac{s_2(t - \tau_2 T_c)}{\sqrt{L_2}} = \sqrt{\frac{P_2}{L_2}} c_I^{(0)}(t - \delta_2 T_c - \tau_2 T_c). \quad (4.8)$$

Thus, the two pilot signals appear to have the same offset at the mobile station if

$$\delta_1 T_c + \tau_1 T_c = \delta_2 T_c + \tau_2 T_c$$
$$\Rightarrow \tau_1 - \tau_2 = \delta_2 - \delta_1 = \delta_{12}, \quad (4.9)$$

where the numbers (in chips) are computed modulo 2^{15}, the period of the pilot PN sequence. The above equation shows that δ_{12}, the path delay difference in chips for cell 1 to interfere with a mobile station being served by cell 2, is not the same as δ_{21}, the path delay difference in chips for cell 2 to interfere with a mobile station being served by cell 1. For example, if $\delta_1 = 1024$ and $\delta_2 = 512$, then $\delta_{21} = 512$ and $\delta_{12} = (-512)$ mod$(2^{15}) = 32256$, and thus cell 2 is much more likely to interfere with cell 1 rather than vice versa.

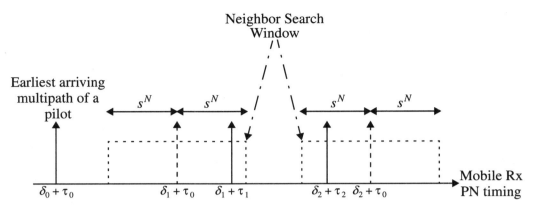

Figure 4–5 Timing Diagram for Criterion S2

Note: Solid arrows denote the actual occurrence of pilot signal, and dotted arrows denote the calculated search window center.

4.4.1.1 Criterion S1: To Prevent the Presence of a Pilot Signal with a **Different** *PN Offset in the Active Search Window Due to a Large Differential Delay*

Figure 4–4 shows a timing diagram for a scenario in which the pilot signal with a PN offset different from that of the active set pilot arrives at the outside of the active search window even when there is a significant differential delay. To ensure that this happens, the following must be satisfied:

$$(\delta_2 + \tau_2) - (\delta_1 + \tau_1) > s_1^A. \tag{4.10}$$

The lower bound for the PN offset difference can then be obtained from

$$\delta_{12} = \delta_2 - \delta_1 > s_1^A + \max\{\tau_1 - \tau_2\}. \tag{4.11}$$

In practice, the delay τ_1 would be the maximum when the mobile is at the cell edge of base station 1. Note in this case that $\max\{\tau_1 - \tau_2\} = r_1$.

4.4.1.2 Criterion S2: To Prevent the Presence of a Pilot Signal with an **Undesired** *PN Offset in the Neighbor Search Window Due to a Large Differential Delay*

In this section, we assume that the pilot offsets δ_1 and δ_2 are for the pilots in the neighbor set when the serving sector transmits a pilot with offset δ_0, and we will derive a bound for δ_{12}.

Figure 4–5 shows a diagram of the mobile receive PN timing. As shown in the figure, the mobile will place the search windows for the neighbor pilots according to the

PN offset and time delay associated with the active pilot. Then the neighbor search windows for δ_1 and δ_2 are centered at $\delta_1 + \tau_0$ and $\delta_2 + \tau_0$, respectively, with a common window width of s^N.

To avoid any confusion about the neighbor search, the two neighbor search windows should not overlap. This condition can be satisfied if

$$\delta_2 + \tau_0 - s^N > \delta_1 + \tau_0 + s^N, \tag{4.12}$$

which reduces to

$$\delta_{12} = \delta_2 - \delta_1 > 2s^N. \tag{4.13}$$

By using (4.11) and (4.13), the lower limit of δ_{12} that satisfies both criteria S1 and S2 can be found from

$$\delta_{12} = \delta_2 - \delta_1 > \max\{s_1^A + r_1, 2s^N\}. \tag{4.14}$$

In the case of $s^N > s_1^A$ and $s^N > r_1$, which would be typical for most of the practical systems, (4.14) becomes

$$\delta_{12} > 2s^N. \tag{4.15}$$

Since $\delta_{12} = PILOT\_INC \times 64$, the lower limit of $PILOT\_INC$ now can be found from

$$PILOT\_INC \times 64 > 2s^N \tag{4.16}$$

or

$$PILOT\_INC > \frac{2s^N}{64}. \tag{4.17}$$

Table 4–1 illustrates the lower limit of *PILOT_INC* for a number of different values of the neighbor search window size.

Use of a smaller value of *PILOT_INC* results in more phase offsets to choose from. In practice, however, a larger value of *PILOT_INC* may be used for the system to provide an extra margin or to reduce the number of pilots in the remaining set. In any case, it is likely (given the current size and growth rates of cellular systems) that the total number of different pilot PN sequence phase offsets available is smaller than the number of sectors in the system. It may therefore be necessary to assign a pilot PN sequence phase offset to more than one sector.

Table 4–1 Lower Limit of *PILOT_INC*

SRCH_WIN_N	$2s^N$ (chips)	Minimum PILOT_INC	Number of PN Offset Indexes*
7	40	1	512
8	60	1	512
9	80	2	256
10	100	2	256
11	130	3	170
12	160	3	170
13	226	4	128

* The offset index is specified by an integer from 0 through 511 inclusive, resulting in a total of 512 distinct pilot channels [4.1] [4.2]. Therefore, given a pilot PN sequence offset index increment *PILOT_INC*, the total number of distinct offset indexes separated by at least *PILOT_INC* × 64 chips is given by $\left\lfloor \dfrac{512}{PILOT\_INC} \right\rfloor$, where $\lfloor x \rfloor$ indicates the largest integer not exceeding x.

The next section outlines some considerations underlying the reuse of pilot PN sequence phase offsets in a cellular system. Note that a smaller value of *PILOT_INC* would result in a larger reuse distance.

4.4.2 Reusing Pilot PN Sequence Phase Offsets

The following two situations may occur when two or more sectors use the same pilot PN sequence phase offset:

- a mobile station being served by one of the sectors can experience interference from the CDMA carrier of the other sector(s) using the same pilot offset, and
- a sector in the system may be unable to uniquely identify all the pilot signals reported by a mobile station that it is serving.

As shown below, providing enough geographical separation between the sectors using the same pilot PN sequence phase offset avoids both of these situations. Specifically, we will consider the situation shown in Figure 4–6 in which the pilot PN sequence phase offsets of the base stations in cell 1 and cell 3 are the same. Let

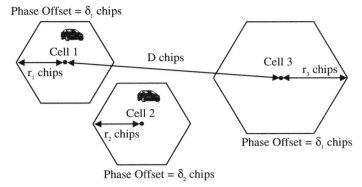

Figure 4–6 Reusing Pilot PN Sequence Phase Offsets

- D denote the distance (in chips) between cell 1 and cell 3,
- r_i denote the radius of the base station i,
- δ_i denote the pilot PN phase offset associated with the base station i,
- τ_i denote the time delay from the base station i to the mobile,
- s_i^A denote the (one-sided) active search window size of the base station i,
- s_2 denote the larger of the (one-sided) neighbor set search window size and remaining set search window size of cell 2,
- γ denote the path loss exponent, and
- P_i denote the transmit powers of the pilot signals of the base station i.

In the following, we derive the lower bound for D according to the criteria listed in Section 4.4.

4.4.2.1 *Criterion R1: To Prevent Undesired Finger Output for the Pilot Signal from Distant Reuse Cell*

Let L_i be the net path loss (including shadow fading and antenna gain) from cell i to the mobile. Following the discussion in [4.3], we have

$$L_i = d_i^\gamma 10^{(a\xi + b\xi_i)/10} = d_i^\gamma e^{\beta(a\xi + b\xi_i)}, \qquad (4.18)$$

where d_i is the distance from cell i to the mobile, $\beta = (\ln 10)/10$, $a = b = 1/\sqrt{2}$, and ξ and ξ_i are the Gaussian random variables representing log-normal shadow fading related to the mobile and cell i, respectively.

As in Figure 4–6, with cell 1 being the serving cell and cell 3 being the distant cell using the same PN offset, criterion R1 can be formulated as

$$\frac{\left(\dfrac{P_1}{L_1}\right)}{\left(\dfrac{P_3}{L_3}\right)} = \frac{P_1 L_3}{P_3 L_1} = \frac{P_1}{P_3}\left(\frac{d_3}{d_1}\right)^{\gamma} e^{\beta b(\xi_3 - \xi_1)} > T \tag{4.19}$$

or

$$\frac{P_1}{P_3}\left(\frac{d_3}{d_1}\right)^{\gamma} > T \bullet e^{\beta b(\xi_1 - \xi_3)} \tag{4.20}$$

where T is the threshold for the ratio of the desired to undesired finger output of a Rake receiver for the pilot signal and should be set such that the undesired finger will statistically never be detected by the mobile.

The worst case would occur when the ratio given in (4.20) is the minimum. The left-hand side of the inequality of (4.20) will be minimum for minimum d_3 and maximum d_1, which corresponds to the case in which the mobile is located at the edge of the serving cell while it is on the straight line connecting the two base stations and has a direct path from the distant cell. Therefore, the condition for the worst case for the ratio in (4.20) is given by

$$\begin{aligned} d_1 &= r_1 \\ d_3 &= D - r_1 \end{aligned} \tag{4.21}$$

Then criterion R1 can be rewritten as

$$\frac{P_1}{P_3}\left(\frac{D - r_1}{r_1}\right)^{\gamma} > T \bullet e^{\beta b(\xi_1 - \xi_3)}. \tag{4.22}$$

Rearranging (4.22) in terms of the reuse distance D gives

$$D > r_1\left(1 + \left(\frac{P_3}{P_1} \bullet T \bullet x\right)^{\frac{1}{\gamma}}\right) \tag{4.23}$$

where

$$x = e^{\beta b(\xi_1 - \xi_3)}. \tag{4.24}$$

To find D satisfying (4.23) for the worst case, we need to know the maximum of the right-hand side of (4.23). Because x takes on values of a log-normally distributed random variable, it is unbounded. For practical purposes, we choose a constant x_{90} such that

$$\Pr\{x \le x_{90}\} = 0.9. \tag{4.25}$$

That is, x is less than x_{90} 90 percent of the time, and we consider x_{90} as the worst case number.[4] In this case, (4.23) becomes

$$D > r_1 \left(1 + \left(\frac{P_3}{P_1} \bullet T \bullet x_{90}\right)^{\frac{1}{\gamma}}\right). \tag{4.26}$$

With $\gamma = 3.84$, $T = 19$ dB (actually $T = 10^{1.9}$), and the standard deviation for the shadow fading of 8 dB, we have

$$D > r_1 \left(1 + \left(\frac{P_3}{P_1}\right)^{\frac{1}{\gamma}} 5.8\right). \tag{4.27}$$

For systems of equal size cells, the transmit power for the pilot signal would be the same for all base stations. Thus, in this case, (4.27) can be further simplified as

$$D > 6.8 r_1. \tag{4.28}$$

Equations (4.27) and (4.28) provide a rule as to the reuse distance required to meet criterion R1.

For a three-sector cell system, we consider Figure 4–7 and modify the above analysis such that P_1 represents the pilot power transmitted by the alpha sector of the serving cell (that is, cell 1) and P_3 denotes the pilot power from the alpha sector transmitter of the distant cell (that is, cell 3). When assuming equal size cells of radius r, (4.26) is rewritten by

$$D > r \left(1 + \left(\frac{P_3 G_b}{P_1 G_f} \bullet T \bullet x_{90}\right)^{\frac{1}{\gamma}}\right), \tag{4.29}$$

where G_f is the front lobe antenna gain and G_b is the back lobe antenna gain.

---

4. This is a concept similar to ninety percent coverage probability that is discussed in Chapter 7. $x_{90} \approx 1.29\sigma$, where σ is the standard deviation of the shadow fading component ξ_i.

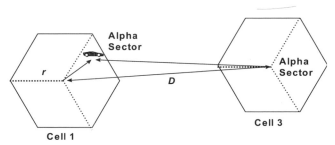

Figure 4–7 Model for Criterion R1 for Three-Sector Cell Case

For an antenna with its 3 dB gain angle in the range of 60 to 100 degrees, the ratio of the front lobe gain to back lobe gain (G_f/G_b) should be at least 10 dB. In this case, with $\gamma = 3.84$, $T = 19$ dB, and 8 dB standard deviation for shadow fading, (4.29) becomes

$$D > 3.2r. \quad (4.30)$$

Note that the use of a three-sector cell layout relaxes the requirement in (4.28). That is, the reuse distance for three-sector cells can be smaller than that required for an omni-cell configuration.

4.4.2.2 Criterion R2: To Guarantee the Absence of the Undesired Finger Output for the Pilot Signal from Distant Reuse Cell

A mobile station in cell 1 can experience interference from the CDMA carrier from cell 3 if the pilot from cell 3 falls within the active search window of the mobile station. This situation can be avoided if

$$\tau_3 - \tau_1 > s_1^A. \quad (4.31)$$

This inequality has to be satisfied even when $(\tau_3 - \tau_1)$ is minimum, which occurs when the mobile station is located at the edge of cell 1 on the straight line connecting the two cells 1 and 3. In this case, $\tau_1 = r_1$, $\tau_3 = D - r_1$, and (4.31) becomes

$$D > 2r_1 + s_1^A. \quad (4.32)$$

Similarly, to avoid interference caused by the CDMA carrier from cell 1 to a mobile station in cell 3, the distance between the two base stations should satisfy

$$D > 2r_3 + s_3^A. \quad (4.33)$$

For equal size cells with the same size of the active search window, that is, for $r_1 = r_3 = r$, $s_1^A = s_3^A = s^A$, (4.32) and (4.33) reduce to

$$D > 2r + s^A. \tag{4.34}$$

Example: Suppose that $S^A = 20$ chips and $r = 2$ miles ($\Rightarrow$ 13 chips). Since $S^A \approx 1.53r$, we get from (4.34)

$$D \geq 3.53r. \qquad \square$$

4.4.2.3 Criterion R3: To Prevent Indistinguishability of Sectors with the Same Offset in Other's Neighbor Search Window

For cell 2 (in Figure 4–6) to uniquely identify the pilot signals reported by the mobile stations that it is serving, cells 1 and 3 should be arranged so that, at most, one of the base stations in cells 1 and 3 is within hearing distance of a mobile station in cell 2 and so that this identity does not change if the mobile station moves to a different position in cell 2. Therefore, at least one of the base stations in cells 1 or 3 must be more than

$$2r_2 + s_2 \text{ chips}$$

away from the base station in cell 2. This condition is satisfied if the distance between the base stations in cells 1 and 3 satisfies (when considering the worst case of the three base stations being on a straight line)

$$D > 2(2r_2 + s_2) = 4r_2 + 2s_2. \tag{4.35}$$

That is, (4.35) is the *sufficient* condition of either the cell 1 or 3 base station being at least $2r_2 + s_2$ chips away from the cell 2 base station. For the case of equal-size cells of radius r where the neighbor set search window is larger than both the remaining set search window and the active set search window, which is typically the case, (4.35) provides a bound tighter than (4.34). That is, the reuse distance should satisfy the inequality

$$D > 4r + 2s^N. \tag{4.36}$$

When the window size is very large (for example, $s^N \gg r$), we consider criterion R3 in terms of the pilot power received at the mobile as we did in 4.4.2.1. To prevent indistinguishability, the interfering signal must be below the detection threshold. That is,

$$\frac{\left(\dfrac{P_1}{L_1}\right)}{\left(\dfrac{P_2}{L_2}\right)} < T_d \text{ or } \frac{\left(\dfrac{P_3}{L_3}\right)}{\left(\dfrac{P_2}{L_2}\right)} < T_d \tag{4.37}$$

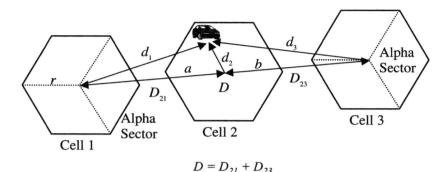

Figure 4–8 Model for Criterion R3 for Three-Sector Cell Case

where T_d is the detection threshold for pilot E_c/I_o. In the above expression, we have assumed that the mobile is located in cell 2 as shown in Figure 4–8. By following an analysis similar to the one in Section 4.4.2.1, the worst case in Figure 4–8 leads to[5]

$$D_{21} > r\left(\left(\frac{1}{T_d} \cdot x_{90}\right)^{\frac{1}{\gamma}} + 1\right) \text{ or } D_{23} > r\left(\left(\frac{G_b}{G_f} \cdot \frac{1}{T_d} \cdot x_{90}\right)^{\frac{1}{\gamma}} - 1\right) \quad (4.38)$$

if the mobile is located at point a or

$$D_{21} > r\left(\left(\frac{1}{T_d} \cdot x_{90}\right)^{\frac{1}{\gamma}} - 1\right) \text{ or } D_{23} > r\left(\left(\frac{G_b}{G_f} \cdot \frac{1}{T_d} \cdot x_{90}\right)^{\frac{1}{\gamma}} + 1\right) \quad (4.39)$$

if the mobile is located at point b.

From (4.38) and (4.39), we conclude that it is sufficient for the reuse distance to satisfy the following expression:

$$D > r\left(1 + \left(\frac{G_b}{G_f}\right)^{\frac{1}{\gamma}}\right)\left(\frac{1}{T_d} \cdot x_{90}\right)^{\frac{1}{\gamma}}. \quad (4.40)$$

---

5. It is assumed here that the alpha sector is the neighbor sector, and the derivation focuses on the alpha sectors of cell 1 and 3. Because of the symmetry, the derivation based on other sectors would not change the conclusion.

With $(G_f/G_b) = 10$ dB, $\gamma = 3.84$, $T_d = -19$ dB, and 8 dB standard deviation for shadow fading, (4.40) results in

$$D > 9r. \quad (4.41)$$

When comparing (4.41) with (4.30), it is obvious that (4.41) is the limiting constraint for the three-sector configuration.

4.4.3 An Upper Limit for *PILOT_INC*

As in the frequency reuse for the analog systems, the set of available PN offsets is assigned to a cluster of cells that forms a reuse pattern. Reference [4.4] describes the relationship of the reuse distance, D, to the cluster size, N, as

$$D = r\sqrt{3N}. \quad (4.42)$$

For the reuse distance given by (4.42) to satisfy the condition of (4.41), we find

$$N \geq 27. \quad (4.43)$$

In particular, for a three-sector cell network, the total number of available pilot PN offsets should be greater than the number of sectors in one cluster, that is,

$$\left\lfloor \frac{512}{PILOT\_INC} \right\rfloor > 3N > 81. \quad (4.44)$$

The above expression provides an upper limit for *PILOT_INC* whereas (4.17) gives a lower bound. More specifically, the upper bound in this case is given by

$$PILOT\_INC \leq 6. \quad (4.45)$$

Note that (4.45) is satisfied for the search window sizes shown in Table 4–1.

4.5 A Procedure for Phase Assignment

The following approach is suggested for assigning pilot PN sequence phase offsets in a cellular system.

1. Setting *PILOT_INC* for the system:
   A lower limit for *PILOT_INC* can be computed using the method outlined in Section 4.4. A better propagation environment or a larger margin (T) would lead to a higher value.

A value of four for *PILOT_INC* is recommended. This value reduces the number of pilot offsets in the remaining set compared to a smaller value of *PILOT_INC* but can also reduce the search time for pilots. Also, with *PILOT_INC* = 4, the RF engineer has a set of 128 different pilot PN sequence phase offsets, which, with appropriate reuse, is large enough to cover most cellular systems. Note that if *PILOT_INC* = 4, then any two different phase offsets in the system are separated by at least 256 chips, which is (approximately) equivalent to forty miles. Therefore, a CDMA carrier would have to travel approximately forty miles before it could cause interference by falling in the active search window of a mobile station synchronized to a different carrier. Under most conditions, the *wrong* CDMA carrier would be weak enough to avoid causing any problems.

2. Assigning pilot PN sequence phase offsets to sectors:

   Once *PILOT_INC* is set, the RF engineer has $\lfloor 512/PILOT\_INC \rfloor$ different offsets for the system. When assigning the offsets to sectors, the following points should be considered.

   - Phase offsets of neighboring sectors should be separated from each other to aid initial pilot signal acquisition in regions where more than one pilot signal can be detected.
   - Different sectors using the same pilot offset should be geographically separated as discussed in Section 4.4.2.
   - Some offsets should be kept reserved for growth or microcells and not be used for the phase assignment plan of the regular grid.

As an example, Figure 4–9 shows a phase assignment in a regular (three-sectored) grid. The assignment is based on *PILOT_INC* = 4, so that there are 128 different pilot offsets. Out of these, 111 pilot offsets are used for assignment in the regular grid, leaving seventeen phase offsets for growth or for cells that do not follow the regular pattern (for example, microcells). In this case, we can see from (4.43) that a cluster of thirty-seven cells can be formed for the purpose of reusing PN offsets, which corresponds to a three-tier system. Note that the reuse distance between two cells that use the same phase offset is more than $10r$, where r is the nominal radius of the cells.

Every cell shown in the figure is sectored into three sectors. A cell that is numbered n is assigned the three phase offsets

$$n \times 256, (n + 43) \times 256, (n + 86) \times 256,$$

A Procedure for Phase Assignment

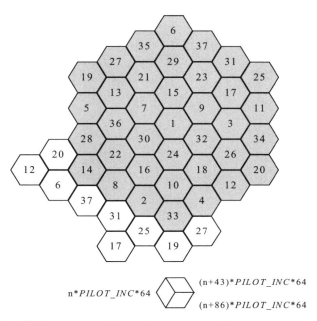

Figure 4–9 Pilot PN Sequence Phase Assignment Plan
Note: The cluster size is thirty-seven, and the recommended value of *PILOT_INC* is four.

which are assigned to the three sectors as shown in Figure 4–9. The shaded region in the figure is the basic building block of the grid. The following seventeen phase offsets are unused in the regular grid and reserved for growth or for cells that do not follow the grid:

$$38 \times 640, 39 \times 640, 40 \times 640, 41 \times 640, 42 \times 640, 43 \times 640,$$

$$81 \times 640, 82 \times 640, 83 \times 640, 84 \times 640, 85 \times 640, 86 \times 640,$$

$$124 \times 640, 125 \times 640, 126 \times 640, 127 \times 640, 0 \times 640.$$

This phase assignment provides a reasonable separation between the pilot PN sequence phase offsets of bordering sectors as explained below.

The 128 different phase offsets available in the system are

$$\{0 \times 256, 1 \times 256, ..., 127 \times 256\}.$$

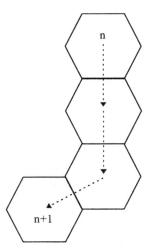

Figure 4–10 The Rule for Assigning a Cell Number Within a Thirty-Seven-Cell Cluster

We divide this set of offsets into the following four sets:

- Set I: $\{1 \times 256, 2 \times 256, ..., 37 \times 256\}$,
- Set II: $\{44 \times 256, 45 \times 256, ..., 80 \times 256\}$,
- Set III: $\{87 \times 640, 88 \times 640, ..., 123 \times 640\}$, and
- Set IV: consists of the 17 remaining offsets.

We use the pilot offsets in Sets I to III for the regular grid and reserve those in Set IV for cells that do not follow the grid. Each cell (in the grid) is assigned one pilot offset from each of the first three sets. We assign the three offsets to the three sectors in a regular fashion as shown in Figure 4–9. With this assignment, we always assign pilot offsets to the four neighboring sectors for any sector (in the grid) from sets other than that used for the sector, thus assuring a reasonable separation.

The thirty-seven cells in the cluster are numbered from one to thirty-seven per the following rule.[6]

> Given a cell n, the next cell number $n + 1$ is assigned to the cell located to the south of cell n by two cells and then to the southwest (60 degrees away from south) by one cell as shown in Figure 4–10. If cell $n + 1$ is outside of

---

6. Although other assign rules may produce better *maximum separation* properties, we believe that the improvement will be marginal. Therefore we have made no attempt here to find the best assignment rule to yield the optimal maximum separation layout.

the three-tier system, then we use the cell within the three-tier system according to the reuse pattern.

Note that the cluster pattern shown in Figure 4–9, which results from this rule, has the following properties.

1. Any first-tier neighbor cells are of at least 6×256 chips separation.
2. Any second-tier neighbor cells are of at least 2×256 chips separation.

4.6 References

[4.1] TIA/EIA/IS-95A. *Mobile Station-Base Station Compatibility Standard for Dual-Mode Wideband Spread Spectrum Cellular System,* March 1995.

[4.2] ANSI J-STD-008. *Mobile Station - Base Station Compatibility Requirements for 1.8 and 2.0 GHz CDMA PCS,* March 1995.

[4.3] A. J. Viterbi. *CDMA Principles of Spread Spectrum Communication*, Addison-Wesley, 1995.

[4.4] V. H. MacDonald. The Cellular Concept, *The Bell System Technical Journal,* vol. 58, pp.15-41, January 1979.

CHAPTER 5

Mobile Station Access and Paging

CDMA mobile stations transmit on the access channel according to a random access protocol. Detailed procedures of this random access protocol and ranges of various access parameters can be found in References [5.1] and [5.2].

This chapter, which consists of four sections, analyzes the protocol performance and discusses appropriate settings of mobile station access parameters. Section 5.1 describes the operation of the mobile station access protocol and its associated parameters. Section 5.2 presents persistence delays for access request attempt. Section 5.3 presents throughput and delay performance of the mobile station access protocol and discusses an approach to determine the access channel capacity. Section 5.4 evaluates paging channel capacity in terms of various paging messages related to call processing and special services such as short message services (SMSs) and voice mail services (VMSs).

5.1 Description of Mobile Station Access Protocol

As described in Reference [5.1], the mobile station transmits on the access channel using a random access procedure. Figure 5–1 shows a flowchart of the CDMA mobile station access protocol. The entire process of sending one message and receiving (or failing to receive) an acknowledgment for that message is called an *access attempt*. Each transmission in the access attempt is called an *access probe* (see Figures 5–2, 5–3, and 5–4). The mobile station transmits the same message in each access probe in an

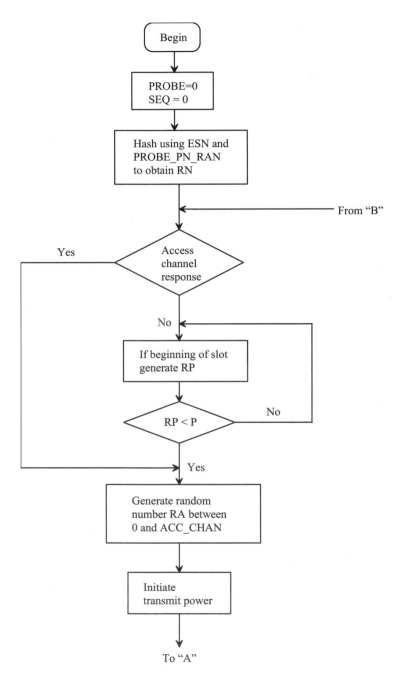

Figure 5–1 Access Procedure

Description of Mobile Station Access Protocol

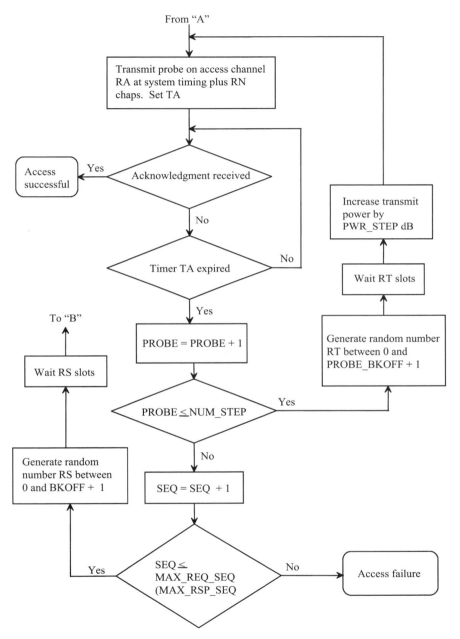

Figure 5–1 Access Procedure (Continued)

Reproduced from [5.1] and [5.2] under written permission of the copyright holder (Telecommunications Industry Association)

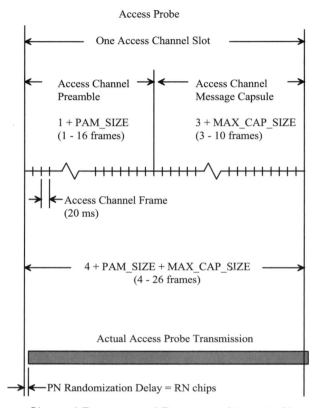

Figure 5–2 Access Channel Request and Response Attempts (Access Probe)
Reproduced from [5.1] and [5.2] under written permission of the copyright holder (Telecommunications Industry Association)

access attempt. Each access probe consists of an access channel preamble and an access channel message capsule. The number of twenty millisecond frames expresses the length of the preamble $1 + pam\_sz$ as well as the length of message capsule $3 + max\_cap\_sz$. Thus, the duration of an access probe (access channel slot) is $4 + pam\_sz + max\_cap\_sz$ frames.

Within an access attempt, access probes are grouped into an access probe sequence. Mobile stations send two types of messages on the access channel: a response message (one that is a response to a base station message, see Figure 5–3) or a request message (one that the mobile station sends autonomously, see Figure 5–4). Each access attempt consists of up to max_req_seq (for a request access) or max_rsp_seq (for a response access) access probe sequences.

Description of Mobile Station Access Protocol

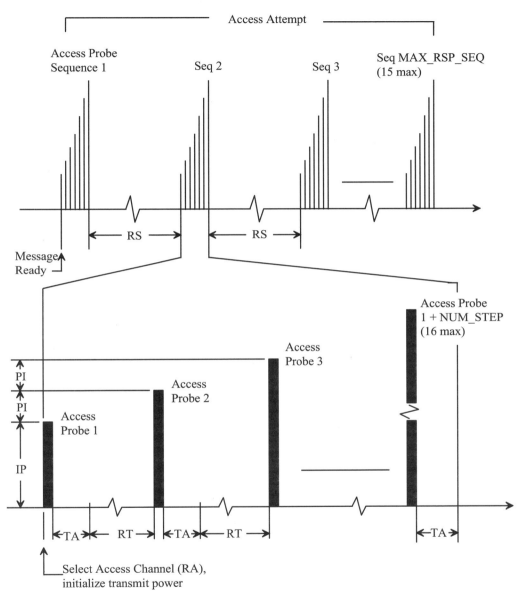

Figure 5–3 Access Channel Response Attempts

Reproduced from [5.1] and [5.2] under written permission of the copyright holder (Telecommunications Industry Association)

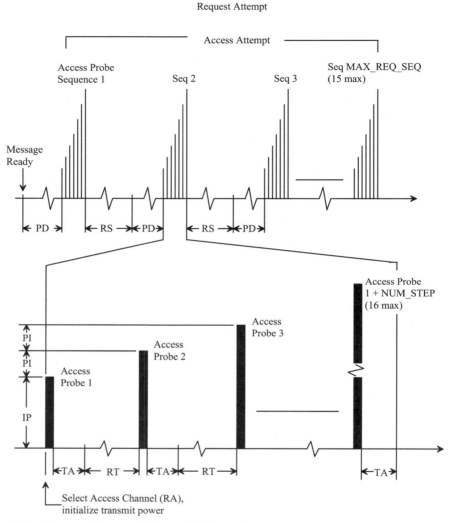

Figure 5–4 Access Channel Request Attempts

Reproduced from [5.1] and [5.2] under written permission of the copyright holder (Telecommunications Industry Association)

Mobile stations pseudo-randomly determine the timing of the start of each access probe sequence. For every access probe sequence, a backoff delay, *RS*, from 1 to 1 + *bkoff* slots is generated pseudo-randomly.

In the case for request access probe sequences, for each slot after the backoff delay *RS*, the mobile station performs a pseudo-random persistence test. If the persistence test passes, the first probe of the sequence begins in that slot. Failure of the persistence test defers the access probe sequence until at least the next slot. Thus, the use of a persistence test imposes an additional delay, *PD*. For each access channel slot, the persistence test generates a random number and compares it with a predetermined threshold. The pre-computed threshold is different, depending on the nature of the request, the access overload class *n*, its persistence value *psist(n)*, and its persistence modifier *msg_psist*, for message transmission or *reg_psist* for registrations.

Each access probe sequence consists of up to 1 + *num_step* access probes, all transmitted on the same access channel. Mobile stations pseudo-randomly choose the access channel number *RA* (used for each access probe sequence) from 0 to *acc_chan* among all the access channels associated with the current paging channel. The mobile station shall use this access channel number for all access probes in that access probe sequence.

The mobile station transmits the first probe in each access probe sequence at a mean output power level (referenced to the nominal CDMA channel bandwidth of 1.23 MHz) depending on the open-loop power estimate, the initial power offset for access *init_pwr*, and the nominal transmit power offset *nom_pwr*. Each subsequent access probe is transmitted at a power level of a specified amount *PI* (determined from *pwr_step*) higher than the previous access probe until it obtains an acknowledgment response or the sequence ends. Between access probes, the mobile station shall disable its transmitter.

Expression of the timing of access probes and access probe sequences is in terms of access channel slots. The transmission of an access probe begins at the start of an access channel slot. A procedure called *PN randomization* determines the precise timing of the access channel transmissions in an access attempt. For the duration of each access attempt, the mobile station computes a delay, *RN*, from 0 to $2^{probe\_pn\_ran}$ PN chips using a (non-random) hash function that depends on its electronic serial number *ESN*. The mobile station delays its transmit timing by *RN* PN chips. This transmit timing adjustment includes delay of the direct sequence spreading long code and of the quadrature spreading I and Q pilot PN sequence, so it effectively increases the apparent range from the mobile station to the base station. This increases the probability that the base station will be able to separately demodulate transmissions from multiple mobile stations in the same access channel slot, especially when many mobile stations are at a similar range from the base station.

The mobile stations also pseudo-randomly generate timing between access probes of an access probe sequence. After transmitting each access probe, the mobile station waits a specified period, $TA = 80 \times (2 + acc\_tmo)$ milliseconds, from the end of the slot to receive an acknowledgment from the base station. If it receives an acknowledgment, the access attempt ends. If not, the mobile station transmits the next access probe after an additional backoff delay RT from 1 to 1 + *probe_bkoff* slots.

5.2 Average Persistence Delay for Access Request Attempt

As discussed in the previous section, for access due to mobile station request, ANSI J-STD-008 [5.1] requires that a persistence test be performed prior to initiating the access probe sequence to control the rate at which the mobile station transmits requests. Assessing the appropriate range of persistence values to be assigned to the mobile stations requires information about the amount of delays due to persistence tests. This section includes calculations of average persistence delays as a function of persistence values for various types of request and access overload classes.

For each access channel slot, the persistence test generates a random number RP ($0 < RP < 1$) and compares it with a predetermined threshold P. The access probe sequence initiates if the generated random number RP is smaller than the predetermined threshold P. Since the random number RP is generated from uniform distribution over the unity interval,

$$\Pr\{RP < P\} = P. \tag{5.1}$$

In other words, a larger P implies a higher probability of initiating the access probe sequence.

The pre-computed threshold P, in general, is different, depending on the nature of the request, the access overload class, and its persistence value *psist(n)* as well as its persistence modifier. As an example, for registration request of access overload classes 0 through 9, if *psist(n) = 63*, then *P = 0*. Thus the persistence test fails and therefore initiates no access probe sequence; if *psist(n)* is not equal to 63, for a given persistence modifier *reg_psist*, P is the monotonically decreasing function of *psist(n)*; the larger the *psist(n)*, the smaller the P, thus the smaller the probability of initiating the access probe sequence. Table 5–1 summarizes the persistence test thresholds for various types of requests and access overload classes.

From Table 5–1, note that the maximum persistence value is 63 for access overload classes 0 through 9, and is 7 for access overload classes 10 through 15. If the base station assigns the maximum persistence value to the mobile station, the access attempts fail and the mobile station enters the *system determination substate* of the *mobile station initialization state* (see References [5.1] and [5.2]).

For the persistence value not equal to the maximum, random persistence delays can be incurred to control the transmissions of mobile station requests. The persistence

Average Persistence Delay for Access Request Attempt

Table 5–1 Persistence Test Thresholds (P) for Registration, Message, and Other Requests

	access overload classes $n = 0, 1, ..., 9$		access overload classes $n = 10, 11, ..., 15$	
	$psist(n) \neq 63$	$psist(n) = 63$	$psist(n) \neq 7$	$psist(n) = 7$
Registration Request	$2^{-\frac{psist(n)}{4} - reg\_psist}$	0	$2^{-psist(n) - reg\_psist}$	0
Message Request	$2^{-\frac{psist(n)}{4} - msg\_psist}$	0	$2^{-psist(n) - msg\_psist}$	0
Other Request	$2^{-\frac{psist(n)}{4}}$	0	$2^{-psist(n)}$	0

delay PD is the number of times the tests have to be performed before the condition $RP < P$ is satisfied. Thus, persistence delay PD is a random variable, and its discrete probability density is geometric with parameter P, that is,

$$\Pr\{PD = k \text{ slots}\} = (1 - P)^k P \qquad \text{for } k = 0, 1, 2, \cdots. \tag{5.2}$$

One performance measure that can characterize the persistence test is the average persistence delay. Let $E\{\cdot\}$ be the expectation operator. The average persistence delay $E\{PD\}$ is

$$E\{PD\} = \sum_{k=0}^{\infty} k \Pr\{PD = k \text{ slots}\} = \frac{1-P}{P}. \tag{5.3}$$

Using the values of persistence test thresholds P in Table 5–1 for various types of request and access overload class, Table 5–2 summarizes their average persistence delay $E\{PD\}$.

We numerically evaluate the derived expressions of the average persistence delay in Table 5–2. For registration requests with persistence modifiers (*reg_psist*) 0 through 7, the average persistence delay as a function of persistence value is plotted for access overload classes 0 through 9 in Figure 5–5, and for access overload classes 10 through 15 in Figure 5–6, respectively. Figures 5–5 and 5–6 can also be used for message requests when persistence modifiers *reg_psist* are replaced by *msg_psist*. For other requests, Figure 5–7 plots the average persistence delay as a function of persistence value for access overload classes 0 through 9. Figure 5–8 plots the average persistence delay as a function of persistence value for access overload classes 10 through 15.

Table 5-2 Average Persistence Delay ($E\{PD\}$) for Registration, Message, and Other Requests

	access overload classes $n = 0, 1, ..., 9$		access overload classes $n = 10, 11, ..., 15$	
	$psist(n) \neq 63$	$psist(n) = 63$	$psist(n) \neq 7$	$psist(n) = 7$
Registration Request	$2^{\frac{psist(n)}{4} + reg\_psist} - 1$	∞	$2^{psist(n) + reg\_psist} - 1$	∞
Message Request	$2^{\frac{psist(n)}{4} + msg\_psist} - 1$	∞	$2^{psist(n) + msg\_psist} - 1$	∞
Other Request	$2^{\frac{psist(n)}{4}} - 1$	∞	$2^{psist(n)} - 1$	∞

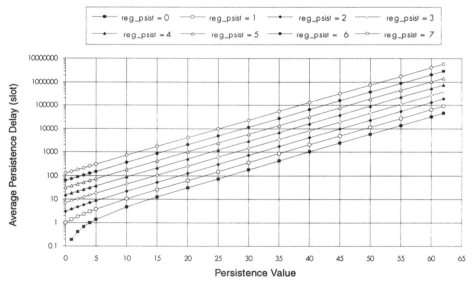

Figure 5-5 Average Persistence Delay for Registration Request of Access Overload Classes 0 through 9

Note: If *msg_psist* is substituted for *reg_psist*, the plot is for the Message Request.

In general, for each type of request and access overload class, maximum allowable persistence delay is expected to be unsurpassed. Thus, given the maximum allowable

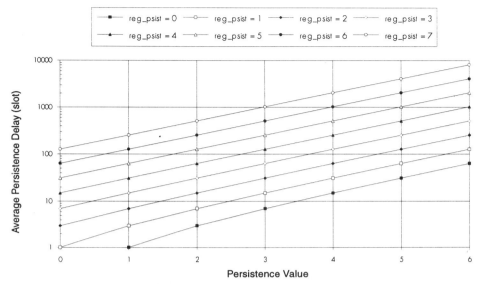

Figure 5–6 Average Persistence Delay for Registration Request of Access Overload Classes 10 through 15

Note: If *msg_psist* is substituted for *reg_psist*, the plot is for the Message Request.

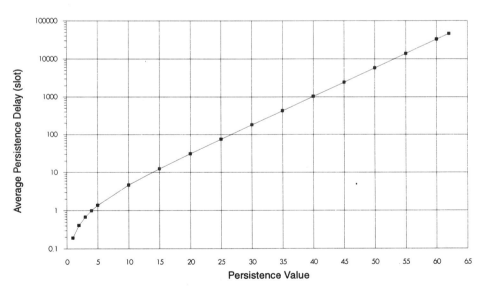

Figure 5–7 Average Persistence Delay for Other Requests of Access Overload Classes 0 through 9

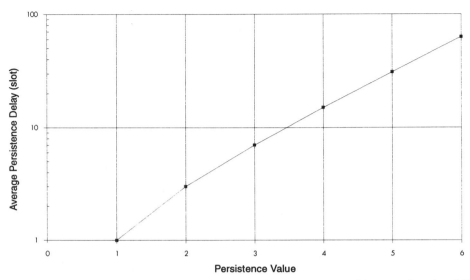

Figure 5–8 Average Persistence Delay for Other Requests of Access Overload Classes 10 through 15

persistence delay for each type of request and access overload class, the appropriate range of persistence values can be assessed.

5.3 Access Channel Capacity

The access channel is used in the reverse link and supports origination calls, page responses, order messages, registrations, and short message services (SMSs).

Even though the access channel uses a slotted random access protocol, a high throughput is possible in the access channel. Each access slot is of a fixed duration and consists of multiple frames, each of which is 20 msec in duration; its long code identifies it uniquely. An access probe is successful if the E_b/I_o at the base station exceeds a threshold[1] and there are no other probes received within an interval of two PN chips. In other words, if two access attempts are separated by an interval of two chips, in most cases, one of the access attempts will get through successfully. Therefore, even when the second access attempt comes before the first access attempt finishes its setup process, the first access attempt can still be successfully decoded without being impacted by the second access attempt.

---

1. The threshold can be dynamic; a fixed signal-to-interference ratio (SIR) is assumed here for simplicity.

Access Channel Capacity

A large amount of access traffic may generate an unacceptable interference in the reverse link. As described in Section 5.1, after the first access probe fails to gain access to the system, the next access probe will increase its probing power by an amount specified by the access channel parameter *PI*. The aggregate effect will result in the increase of the total power received at the cell site and the reduction of the reverse-link capacity. To guarantee the capacity for the voice traffic channels, any excessive use of the capacity by the access channel should be limited.

In this section, we develop an approach to evaluate the IS-95 access channel capacity. After investigating, in Sections 5.3.1 and 5.3.2, the impact of access traffic on the reverse-link interference through a simulation, we estimate the capacity of a single access channel in terms of the number of CDMA carriers that can be supported, which is the subject of Section 5.3.3. The results obtained in this section can be used as a guideline to estimate the capacity limitation and the impacts of other services, such as SMS, on the access channel.

5.3.1 Access Channel Simulation Model

To predict the capacity limit on the access channel, we use a simulation model of the access channel and a criterion based on the interference impact of the access channel on reverse link. This section describes the simulation model used to evaluate access channel performance.

Within each access slot (seven frames for this case; see Table 5–3), we assume the arrival rate of the access attempt to have a Poisson distribution with a given mean. We also assume that the voice activity is 0.4 and the out-of-sector reverse-link interference is seventy-two percent of the in-sector interference. Since we are interested in the capacity of the access channel, we assume that voice traffic channel power is constant and the system is operating at fifty percent loading.[2] The E_b/I_o required for the access probes to be successful is typically higher than that for the voice traffic channel (see, for example, Reference [5.3]).[3] We assume that the required E_b/I_o for an access success is fifty percent higher (or 1.76 dB above) that of the voice traffic channel.

The open-loop power control algorithm determines the initial mobile transmit power used for an access attempt. Because of the imbalance in the RF environment

---

2. Loading is in reference to the pole capacity (or pole point). The pole capacity, as discussed in Section 2.2.1.1, is a theoretical CDMA capacity by assuming infinite mobile transmission power. For example, the pole capacity of a CDMA system using an 8 Kbps vocoder is about forty radio channels. In this case, a fifty percent loading means that the system is operating at a capacity of twenty radio channels. For more discussion of the loading and pole capacity, see Section 8.1.
3. We will define traffic channel E_b/I_o in Chapter 7 (see, for example, (7.5)).

Table 5–3 Access Channel Parameters

Parameter	Value	Note
pam_sz	1 frame	preamble
max_cap_sz	2 frames	information
message(slot) length	7 frames	4+ pam_sz + max_cap_sz
num_step	3	number of probes
max_req_seq and max_rsp_seq	2	

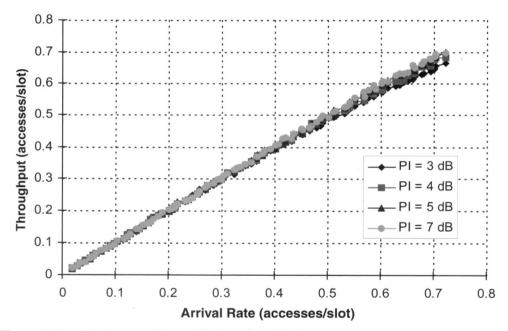

Figure 5–9 The Access Channel Throughput Performance versus Arrival Rate

between the forward- and reverse-links, the imperfections in the open-loop power control algorithm, and the loading in the system, the received signal-to-interference ratio resulting from the first access probe could be different from what is desired. We model the initial access channel power as a Gaussian random variable with mean that is 1.76 dB above the mobile transmit power needed to meet the traffic channel E_b/I_o require-

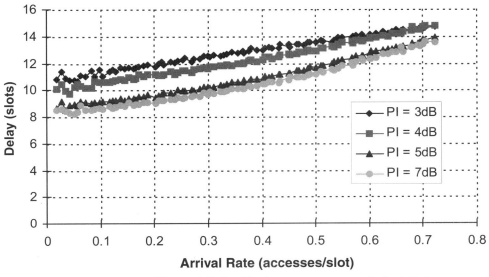

Figure 5–10 The Access Channel Delay Performance versus Arrival Rate

ment and variance of 8 dB. In this simulation, we do not explicitly consider the effect of channel variation (for example, fading) on the access channel.

During the simulation, for a given number of access attempts per slot (or arrival rate), we count the number of successes per slot (or throughput) and the delays experienced by the successful attempts. The time from the completion of the transmission of the first access probe until the time the mobile station receives an acknowledgment defines the delay. We also measure the total received power on the access channel in terms of a fraction of the assumed sector (or cell) loading of fifty percent. For each value of the arrival rate and *PI*, we ran ten thousand trials to get average values of the throughput, delay, and power level. Table 5–3 lists all access channel parameters used in the simulation.

5.3.2 Access Channel Simulation Results

Figures 5–9 to 5–11 summarize some of the simulation results, showing the access channel performance in terms of throughput, delay, and interference impact on reverse link.

Figure 5–9 plots the access channel throughput as a function of the access traffic arrival rate for a number of different values of *PI*. The plot shows that there is only very

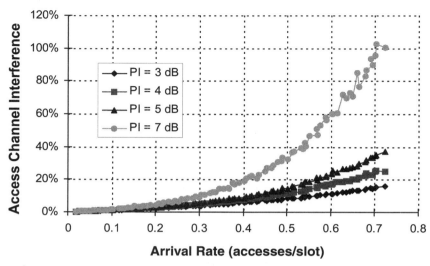

Figure 5–11 The Access Channel Interference Impact versus Arrival Rate

minor improvement by having a large value of *PI* (for example, 7 dB). Basically, the throughput is nearly insensitive to the choice of *PI*.

In Figure 5–10, we plot the delay performance versus the amount of the access channel traffic. The improvement is less than two slots when increasing *PI* from 3 dB to 7 dB. The improvement decreases as the access traffic gets higher.

Next, Figure 5–11 illustrates the amount of reverse-link interference generated by the access attempts. The access channel interference percentage shown in Figure 5–11 is with respect to the fifty percent traffic channel loading assumed in Section 5.3.1. Based on the results from Figure 5–9 and Figure 5–10, the *PI* of 7 dB may be chosen for the best performance of the throughput and delay. However, in Figure 5–11, we find that an excessive interference on the reverse link is possible for a big *PI* when the access traffic is very high. This implies that a smaller *PI* should be considered from the interference viewpoint. In addition, the throughput and delay performance of the access channel is not quite sensitive to the different *PI*. Therefore, a small *PI* should be chosen to minimize the interference impacts. This being the case, we choose *PI* of 3 dB when investigating the access channel capacity in Section 5.3.3.

5.3.3 Access Channel Capacity Analysis

As discussed before, it is important to consider the impact of interference from the access channel in investigating access channel capacity. With a high-access demand, the

Access Channel Capacity

Table 5–4 The Total Number of Subscribers Supported by a Sector*

Number of CDMA RF Carriers	Traffic (Erlangs)	Number of Subscribers (S)
1 (20 radio channels)	13.2	660 (= 13.2/0.02)
2 (40 radio channels)	31	1,550 (= 31/0.02)
3 (60 radio channels)	49.6	2,480 (= 49.6/0.02)
4 (80 radio channels)	68.7	3,435 (= 68.7/0.02)
5 (100 radio channels)	88	4,400 (= 88/0.02)
6 (120 radio channels)	107	5,370 (= 107.4/0.02)

* Trunking efficiency has been considered for Erlang traffic. In a CDMA system, however, the air interface capacity does not necessarily inherit the trunking efficiency by having multiple carriers.

impact on reverse link might not be acceptable because, in this case, the reverse-link capacity decreases too much. Therefore we need to ensure a proper engineering of the access channel to limit the interference impact on reverse-link capacity. In general, we want to keep the access channel interference margin below ten percent of the traffic channel loading.[4] For example, from Figure 5–11, in the case of PI of 3 dB, the interference margin at ten percent of the traffic channel loading corresponds to sixty percent access channel utilization.[5]

5.3.3.1 *Erlang Traffic and Access Channel Traffic Assumptions*

When a single frequency carrier of a CDMA system using an 8 Kbps vocoder can support twenty radio channels, it supports a capacity of 13.2 Erlangs with a two percent blocking probability.[6] Assuming that the user traffic is 0.02 Erlang/subscriber (which is equivalent to each subscriber contributing one seventy-two-second call per hour), we are able to calculate the number of subscribers that a sector can support. By extending this concept to multiple-carrier cases, Table 5–4 shows the number of subscribers that a sector with multiple CDMA RF carriers can support.

As mentioned before, the access channel supports call origination/termination, paging responses, orders, registrations, and SMSs. Statistically, each subscriber will

---

4. With the assumed traffic channel loading of fifty percent, this corresponds to five percent of the pole capacity. In this case, the total sector loading on the reverse link will be fifty-five (= 50 + 5) percent.
5. Note that the arrival rate in accesses/slot is translated into the access channel *utilization* here.
6. This is from the Erlang B table; see, for example, [5.4].

Table 5–5 Access Channel Traffic Assumptions

Access Channel Traffic Type	Value	Note
call origination & termination (OT)	1/hr/subscriber[*]	average number
short message service (SMS)	m/hr/subscriber	variable
time-based registration (Tr)	1 to 2/hr/subscriber	default value
zone-base registration (Zr)	0.1 to 0.2/hr/subscriber	assume 10% of time-based registration
power on/off registration (Pr)	0	cancel out with time-based registration

[*] This is an average number. Some subscribers might make more than one call but others might make no calls during the time of interest.

contribute to the access traffic on different access traffic types. In Table 5–5, we list the values assumed for different access traffic types, which we will need in the next section.

5.3.3.2 Access Channel Capacity Calculation

The number of access messages per hour, A, required to support the access attempts of the types in Table 5–5 is given by

$$A = (OT + SMS + Tr + Zr + Pr) \times S, \quad (5.4)$$

where S is the number of subscribers supported by a sector shown in Table 5–4.

Let M be the access channel capacity. Then, from Table 5–3, we find that the access channel capacity in terms of the number of messages that can be processed by an access channel is given by

$$M = \frac{3600 \text{ sec/hour}}{(20 \times 10^{-3} \text{ sec/frame}) \times (7 \text{ frames/message})} = 25{,}714 \text{ messages/hour}. \quad (5.5)$$

We define the utilization of the access channel as

$$\rho = A/M. \quad (5.6)$$

As discussed in Section 5.3.3, with a limit of a ten percent access channel interference margin, the value of value of ρ should be less than sixty percent. In other words, the access traffic cannot exceed sixty percent of the access channel capacity.

In Figures 5–12 and 5–13, using the numerical values in Table 5–4 and Table 5–5, we plot the access utilization as a function of m, the number of SMS messages. From Figure 5–12, when AR = 30 minutes (one autonomous registration per thirty minutes)

Access Channel Capacity

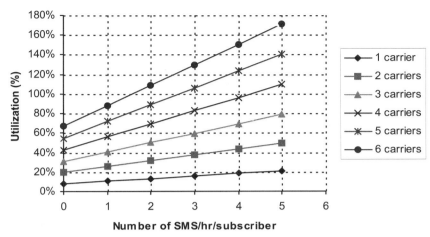

Figure 5–12 Access Channel Utilization with AR = 30 Minutes

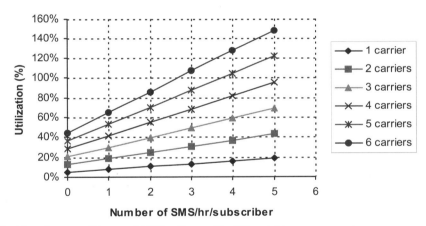

Figure 5–13 Access Channel Utilization with AR = 1 Hour

and SMS = 1/subscriber/hour, one access channel can support four CDMA RF carriers. In Figure 5–13, if we increase the AR time to one hour, then one access channel can support five CDMA RF carriers with SMS = 1/subscriber/hour.

Clearly, higher demand on the SMS and the increased busy hour call attempt (BHCA) will reduce the number of CDMA RF carriers that each access channel can support. It can be shown, however, that even in the case of two SMSs and two call attempts per subscriber during the busy hour, one access channel can still support three CDMA RF carriers when AR = 30 minutes.

We can extend the above results for the 8 Kbps vocoder case to the 13 Kbps vocoder case. In systems using 13 Kbps vocoders, the number of subscribers supported by a single CDMA RF carrier is smaller, thus, users on a single RF carrier will generate less access traffic. Therefore, approximately one access channel of a system using 13 Kbps vocoders can support about one and a half times as many CDMA RF carriers as compared to the systems using 8 Kbps vocoders. Of course, this capacity could vary with different assumptions on the access traffic.

5.4 Paging Channel Capacity

In this section, we discuss an approach to evaluate the capacity of an IS-95A CDMA paging channel. The capacity model developed in this section incorporates most of the salient features of this channel and can be used to quickly answer a variety of questions related to paging channel sizing. We can, for example, calculate the residual capacity available for SMS and voice mail services (VMSs) after accounting for the call-processing load associated with a single paging channel.

In a CDMA system, a paging channel conveys information from base stations to mobile stations. There are three major types of call-processing-related messages.

- The first is an overhead message. It contains information required for call setup (for example, system parameter messages, access parameter messages, neighbor list messages, channel list messages, and extended system parameter messages) and is updated periodically to ensure a successful call setup.
- The second is a page message (or general page message). The page message is used to page the mobile. The page message is sent when a mobile switching center (MSC) receives a call/service request for a mobile. Depending on the paging strategy, the page messages may be sent to a large area through the paging channel on all sectors.
- The third type is a channel assignment message and order message. These messages are important for interacting with a mobile to complete a call/service setup. The base station usually sends these messages only to a small area (a few sectors) during the call/service setup.

In addition to paging messages related to call processing, there are messages associated with supplemental services such as SMS or VMS, which can be sent through the paging channel as well.

In Section 5.4.1, we describe the characteristics of the paging channel that must be considered to formulate a relevant capacity model. In Section 5.4.2, we list all page messages and assumptions used for calculating the paging channel occupancy for each page message type. Section 5.4.3 provides the results for the paging channel capacity and the residual capacity for SMS and VMS after accounting for the call-processing load associated with a single paging channel. In Section 5.4.4, we briefly discuss the results and make some recommendations.

5.4.1 Paging Channel Characteristics

5.4.1.1 *Paging Channel Structure*

As specified in IS-95A [5.2], the data rate for the paging channel can be either 4800 bps or 9600 bps. Unless otherwise specified, we will assume that the paging channel data rate is 9600 bps, but the capacity model to be developed will be applicable to both transmission rates. Figure 5–14 illustrates the paging channel structure.

The paging channel is partitioned into 80 msec paging slots, and each slot consists of eight half frames, each of 10 msec duration. Each half frame begins with a synchronized capsule indicator (SCI) bit, and the first new message in a paging slot[7] must begin immediately following an SCI bit that is set equal to 1. Paging channel messages are carried in paging channel capsules that consist of the message body, an eight-bit length field that indicates the length in bits of the entire capsule, and a CRC code of thirty bits. As we will describe later, paging messages, channel assignments, and orders have lengths in the 100 to 150 bits range, and since a paging slot consists of 760 bits (eight half frames, each having ninety-five potential payload bits), it can potentially carry multiple pages, channel assignments, and orders.

A synchronized paging channel message capsule begins immediately following an SCI bit (in which case the SCI bit is set equal to 1). Since most page, channel assignment, and order message capsules occupy roughly one and a (typically small) fraction of a second half frame, insisting that all message capsules be synchronized would waste a good part of the second half frame. To avoid this inefficient use of a paging channel, the standard permits only the first new message capsules in a slot to be synchronized, and subsequent message capsules in the slot can be appended to the end of the preceding capsule. The message length field in a capsule indicates where the next message capsule in a slot begins. If a message capsule ends less than eight bits from an SCI bit, the standard dictates that the next message in the slot must be synchronized. The length of page and channel assignment capsules must be an integer number of bytes; both of

---

7. Paging channel messages may spill over from a previous slot.

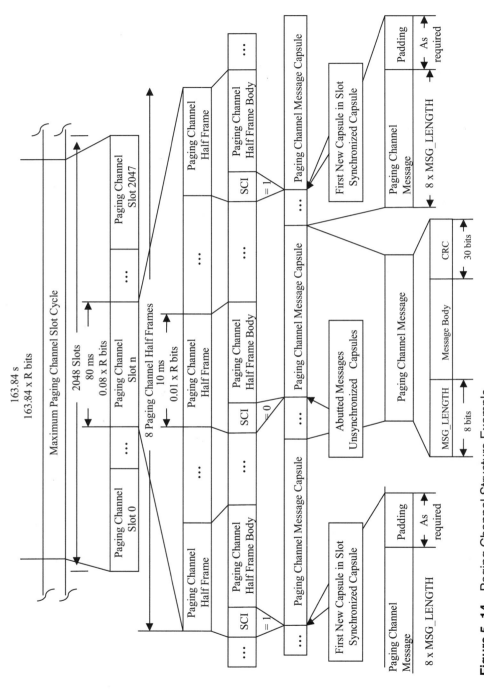

Figure 5-14 Paging Channel Structure Example
Reproduced from [5.1] and [5.2] under written permission of the copyright holder (Telecommunications Industry Association)

these message types have zero to seven bits reserved for padding to satisfy this requirement.

5.4.1.2 Slotted and Non-Slotted Mode

A mobile may operate in non-slotted mode, in which case the mobile reads all page slots while in the *mobile station idle state*. Alternatively, a mobile may (choose to) operate in the slotted mode while in the *mobile station idle state*, in which case the mobile wakes up periodically in specified slots to check for paging messages directed to it. Slotted mode permits the mobile to power down until its pre-specified slot comes along. The mobile wakes up for one or two slots in a slot cycle[8] the length of which can be negotiated between the mobile and the system. The minimum length of a slot cycle is sixteen slots (1.28 sec); for normal cellular/PCS service, this minimum length is likely to be widely used. Mobile stations may use longer slot cycles (maximum length permitted is 2,048 slots), but this would lead to significant delays in terminating a call to mobiles operating in the slotted mode.

Both the mobile and the MSC use the same hash function (see References [5.1] and [5.2]) with the mobiles MIN (or TMSI) as an argument to determine the slot (in the slot cycle) in which the mobile will next awaken. This permits the cellular system to determine the correct slot in which to page mobiles in slotted mode. Mobiles in slotted mode do not generally read the entire paging channel slot in which they awaken because the system sends a _DONE message after all messages scheduled for the slot have been sent (this is an empty general page message with the appropriate field set to indicate that all messages in the slot have already been sent). This permits mobiles in slotted mode to conserve even more battery power.

The requirement that a paging channel message must be contained in two successive slots (the message capsule cannot exceed 1,520 bits) constrains the size of a paging channel message. As indicated earlier, a mobile in slotted mode may read a message that continues onto the next slot; in this instance, the mobile may read the paging channel for as long as two slots.

When IS-95A was approved, it defined three distinct page messages: a slotted page message, a page message, and a general page message. From the viewpoint of paging channel usage, there is little to distinguish among these three distinct page message types because they all have approximately the same length.

---

8. In cases where a page message carries over to a second slot, a mobile would stay awake for the second slot to read the complete message.

Table 5–6 Paging Channel Messages

Message Name
System Parameter Message
Access Parameter Message
Neighbor List Message
CDMA Channel List Message
Slotted Page Message
Page Message
Order Message
Channel Assignment Message
SSD Update Message
Data Burst Message
Authentication Challenge Message
Feature Notification Message
Extended System Parameters Message
General Neighbor List Message
Status Request Message
Service Redirection Message
General Page Message
Global Service Redirection Message
TMSI Assignment Message
Null Message

5.4.2 Assumptions

Table 5–6 lists the twenty paging channel messages that the IS-95A (Reference [5.2]) defines. Of the twenty messages, the slotted page message and the page message are no longer in IS-95B (Reference [5.5]). Of the remaining eighteen messages, the following paging channel messages account for most paging channel usage:

Table 5–7 Assumptions and Message Lengths

	Numerical Values
General Assumptions	
a. Paging Channel Capacity	9600 bits/sec
b. Maximum Allowable Utilization	0.9
c. Paging Strategy (number of pages per users)	1.5
d. Termination Rate	0.35
e. Busy Rate	0.03
f. BHCA per subscriber	2
g. Number of Sectors per MSC	200
h. **General Page Message**	136 bits
i. **Overhead Message**	= j + k + l + m + n
j. System Parameter Message	264 bits
k. Access Parameter Message	184 bits
l. Neighbor List Message	216 bits
m. CDMA Channel List Message	88 bits
n. Extended System Parameter Message	112 bits
o. **Channel Assignment Message**	144 bits
p. **Order Message**	102 bits
Voice Mail Service q. Voice Mail Notification	720 bits
Short Message Services r. Data Burst Message (x = number of characters)	(7x + 380) bits
s. _DONE Message	72 bits

1. General Page Message,
2. Overhead Message,
3. Channel Assignment Message,
4. Order Message, and
5. Data Burst Message for SMS.

The remaining messages are not constantly used. Table 5–7 lists all assumptions and message lengths used for the later calculation.

5.4.3 Paging Channel Capacity

In this section, *we use ninety percent paging channel capacity to be the maximum allowable paging channel utilization*, which is indicated in Table 5–7. We reserve the remaining ten percent to accommodate potential burst paging traffic. The paging channel occupancy for each message is then calculated as a fraction of the maximum allowable paging channel capacity. In other words, if the calculated occupancy is twenty percent, it means that the twenty percent of the ninety percent physical paging channel capacity will be needed to transport the message.

In the following sections, we will evaluate the paging channel occupancy for each message. This will give us a quick answer to a variety of questions related to paging channel sizing. We will use the BHCA as a traffic reference for calculating paging channel capacity.

5.4.3.1 *General Page Message (Call Termination)*

Base stations use the general page message to page (find) a mobile when there is a terminating call or to notify the arrival of a special message service (for example, SMS or VMS). In general, the base station will send the message system-wide to locate the mobile. The paging channel occupancy (with respect to ninety percent paging channel utilization) for the general page message (Og) is calculated as[9]

$$Og = \frac{BHCA \times (d-e) \times c \times h}{3600 \times a \times b}. \tag{5.7}$$

From (5.7), we see that the paging channel occupancy for the general page message is linearly proportional to the growth of traffic demand (that is, BHCA). Figure 5–15 plots the paging channel occupancy for the general page message at different traffic demand. This shows that there is about thirty percent to forty percent occupancy at 150 K to 200 K BHCA. Note that if we can page a small area only (knowing where the mobile is), the reduction of paging channel occupancy can be substantial. For example, if the system pages only one-fourth of the areas, then there is about twenty percent to thirty percent paging channel usage reduction at the range of 150 K to 200 K BHCA traffic.

5.4.3.2 *Overhead Message*

As described in References [5.1] and [5.2], mobile stations need to update all overhead messages to have a successful call setup. Reducing call setup duration can improve successful call setup rate because the probability of losing a strong pilot is pro-

---

9. In equations (5.7) to (5.14), we use item indices in Table 5–7.

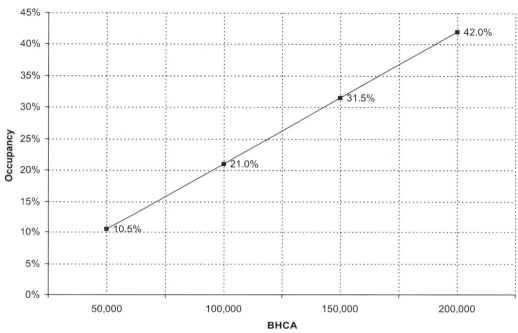

Figure 5-15 Paging Channel Occupancy—General Page Message (Call Termination)

portional to the setup duration. In this section, we will evaluate the impacts of different overhead cycles on the paging channel capacity.

The overhead message (which includes system parameter messages, access parameter messages, neighbor list messages, channel list messages, and extended system parameter messages) will be sent within every n slots[10] to a mobile. The occupancy is independent of the traffic demand. The system will continuously send overhead messages. The paging channel occupancy for the overhead message (Oo) is calculated as

$$Oo = \frac{i \times [1/(0.08 \times \text{Slot Cycle})]}{a \times b}. \quad (5.8)$$

Equation (5.8) indicates that the paging channel occupancy for an overhead message is inversely proportional to the slot cycle duration. Figure 5–16 plots the paging channel occupancy versus varied overhead cycle slots. It shows that the paging channel occupancy grows exponentially with the reduction of the overhead cycle, which suggests that the overhead cycle slots should be greater than seven to avoid using too much

---

10. Each slot is 80 msec.

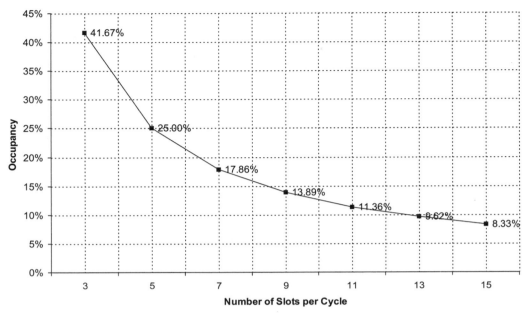

Figure 5–16 Paging Channel Occupancy – Overhead Message

paging channel capacity. Note that reducing the overhead message cycle from fifteen to seven results in approximately a nine and a half percent increase in paging channel usage.

5.4.3.3 *Channel Assignment Message and Order Message*

The base station sends channel assignment messages (CAMs) to the mobile during a call setup (either an originating or a terminating call). Order messages are used for registration reject and base station acknowledgments. The base station transmits CAMs and order messages, interacting during the call/service setup, to the mobile through current communicating sectors' paging channels. The channel assignment message and order message almost come as a pair for each call setup. Since order messages also include registration reject, it has slightly higher rates. For simplicity, we assume that the rates for a channel assignment message and order message are the same. We also assume that the system (one MSC) has two hundred sectors. The paging channel occupancy for channel assignment and order messages, O_{co}, is calculated as

Paging Channel Capacity

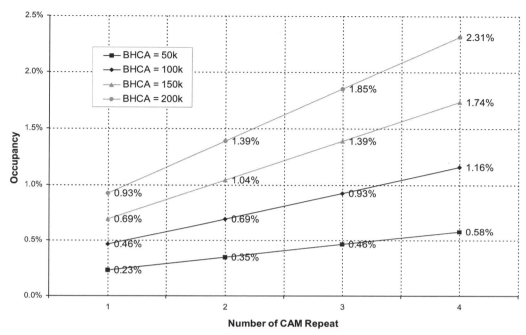

Figure 5-17 Paging Channel Occupancy—Channel Assignment and Order Messages

$$\text{Oco} = \frac{(\text{BHCA}/g) \times (N \times o + p)}{3600 \times a \times b}, \quad (5.9)$$

where N is the number of repeats for channel assignment messages.

The first item on the right-hand side of (5.9) is a simple calculation for the rate of channel assignment and order messages. Clearly, it will tend to underestimate the areas with higher traffic. Figure 5-17 shows the paging channel occupancy for channel assignment messages and order messages with varied CAM repeats and different traffic demands (BHCA). The results show that the paging channel occupancy for channel assignment messages and order messages is low. Even with four repeats and 200 K BHCA, the occupancy is only 2.31 percent.

5.4.3.4 _DONE Message and SCI bit

As mentioned in Section 5.4.1.2, the base station sends the _DONE message after all messages scheduled for the slot have been sent. Here, we assume that there is a _DONE message in each slot. The SCI bit is inserted in each half frame. Therefore, the paging channel occupancy for _DONE message and SCI bits (Ods) can be calculated as

Table 5–8 Paging Channel Occupancy for _DONE Message and SCI bits

Message	Paging Channel Occupancy
_DONE Message (s)	10.42%
SCI	1.16%
Ods	11.58%

$$\text{Ods} = \frac{(1/0.08) \times s + 1 \times (1/0.01)}{a \times b}. \tag{5.10}$$

Table 5–8 lists the results for the paging channel occupancy due to _DONE messages and SCI bits.

5.4.3.5 *Paging Channel Capacity (No SMS and No VMS)*

Figure 5–18 plots the total paging channel occupancy (sum of those given by (5.7)-(5.10)) with different BHCA and overhead message cycles of seven and fifteen. Considering 150 K BHCA, the paging channel occupancy is about fifty-three percent for the overhead message cycle of seven and sixty-three percent for the overhead cycle of fifteen. Therefore, there is about thirty-seven percent to forty-seven percent residual capacity available for the services such as VMS and SMS.

5.4.3.6 *Voice Mail Services (VMSs)*

Voice mail service is a feature that allows callers to leave a voice mail message on a voice mail center. When a caller leaves a voice mail message, the voice mail center will then trigger a request to send a notification to the mobile.

There are several ways to send voice mail notification messages. Here, we will introduce two methods of VMS. The first one is simply using a call setup procedure. The system will send a general page message periodically until the mobile responds. After following the regular call setup procedure, the base station will then send the voice mail through a traffic channel.

The other method uses the SMS type transmission through the paging channel. In this method, no voice mail indication (VMI) message will be sent until the mobile powers on, registers, or sets up a call. The system will first send a general page message, which is followed by a VMI message (through the currently communicating sector's paging channel) upon receiving the response from the mobile.

Let Ov1 denote the paging channel occupancy for the first method and Ov2 denote the paging channel occupancy for the second method. Then,

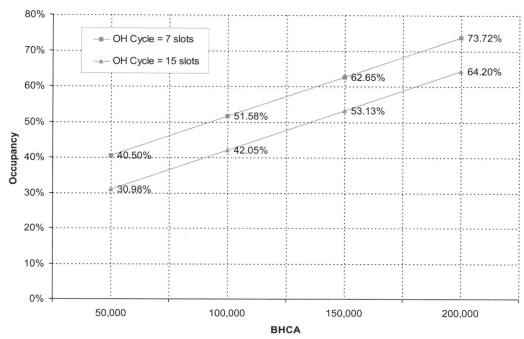

Figure 5–18 Paging Channel Occupancy (No VMS and No SMS)

$$\mathrm{Ov1} = \frac{[(\mathrm{BHCA}/f) \times \mathrm{NOrate} \times 2 \times (d-e) \times (60/v) + \mathrm{BHCA} \times e] \times h}{3600 \times a \times b} \quad (5.11)$$

$$\mathrm{Ov2} = \frac{\mathrm{BHCA} \times e \times (h + q/g)}{3600 \times a \times b}, \quad (5.12)$$

where v is the VMS cycle duration and NOrate is the NO PAGE RESPONSE rate.

Figure 5–19 shows that, using method 1, the paging channel occupancy will substantially increase when the interval between two general page messages, v, is short. With the assumption of twenty percent no page response rate, there is a potential problem when the traffic demand is greater than 150 K BHCA. In this case, for a fifteen-minute page cycle, the paging channel occupancy is close to twenty percent. Most of them are used to notify (page) those mobiles that will never respond.

In method 2, with varied number of sectors in each MSC (that is, g), the SMS transmission can substantially reduce the paging channel occupancy from twenty percent to less than three percent as plotted in Figure 5–20. Note that it is almost insensitive to the number of sectors in one MSC.

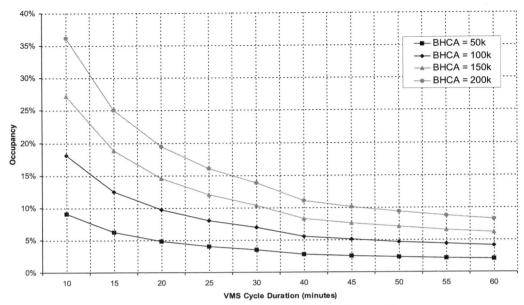

Figure 5–19 Paging Channel Occupancy for VMS (Call Setup Method)

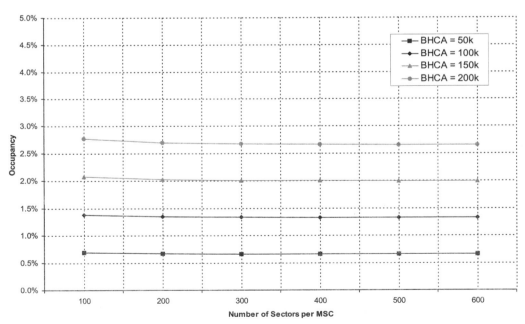

Figure 5–20 Paging Channel Occupancy for VMS (SMS Method)

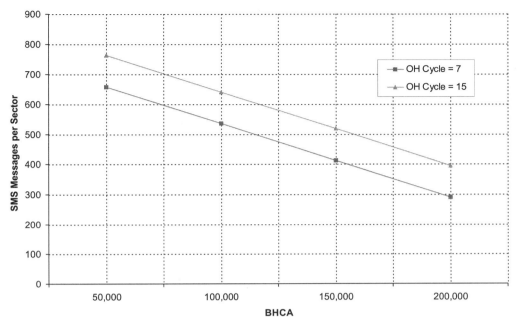

Figure 5–21 SMS Capacity (No VMS)

5.4.3.7 *Data Burst Messages (Short Message Services)*

The base station uses data burst messages to send SMS messages to a mobile. In addition to the seven bits sent for each character in the SMS message (the size of an SMS message in bits is denoted by r; each SMS message contains eighty characters), the data burst message capsule carries approximately 380 bits of overhead. (Of these bits, 232 are due to IS-637 [5.6] mandated overhead, 38 bits are due to the length field and the CRC, and the remaining 110 bits are due to fields in the data burst message). If we assume a flood-paging scheme, every CDMA paging channel in the MSC carries an SMS page for each delivered SMS message. In this case, every paging channel carries $g \times M$ SMS pages/sec. The paging channel occupancy for SMS, Os, can then be calculated as

$$\text{Os} = \frac{M \times (r + h \times g)}{3600 \times a \times b}. \tag{5.13}$$

Given a residual capacity, the number of SMS messages for each sector can be obtained by

$$M = \frac{Os \times (3600 \times a \times b)}{r + h \times g}. \tag{5.14}$$

Figure 5–21 shows the residual capacity for SMS messages (without including VMS). Given that each sector supports about 400 subscribers[11] for a system using 13 Kbps vocoders, the residual paging channel capacity can support about one SMS message per subscriber (at 150 K BHCA) during the busy hour.

5.4.4 Summary

At 150 K BHCA traffic, there is about a fifty-three to sixty-three percent paging channel occupancy used by messages related to call processing. The residual capacity for SMS is about one SMS message per subscriber during the busy hour. The SMS type transmission should be chosen for VMS to avoid unnecessary waste of paging channel capacity. Addition of new services such as SMS and VMS may have caused paging channel usage to reach its maximum allowable capacity.

This situation will be even worse in IS-95B in which the overhead message size is larger and the paging channel loading is higher due to the new access handoff and the soft CAM features. To ensure the appropriate paging channel performance, we need to carefully choose a transmission algorithm for each paging message.

5.5 References

[5.1]   ANSI J-STD-008. *Mobile Station - Base Station Compatibility Requirements for 1.8 and 2.0 GHz CDMA PCS*, March 1995.

[5.2]   EIA/TIA IS-95A. *Mobile Station – Base Station Compatibility Standard for Dual-Mode Wideband Spread Spectrum Cellular System*, March 1995.

[5.3]   A. J. Viterbi. *CDMA: Principles of Spread Spectrum Communication*, Addison-Wesley Publishing Company, 1995.

[5.4]   W.C.Y. Lee. *Mobile Cellular Telecommunications Systems*, McGraw-Hill, 1989.

---

11. For a system using 13 Kbps vocoders, the Erlang traffic is about 7.4. If we assume that each call holding time is seventy-two seconds, the number of subscribers for each sector is 7.4/0.02 = 370.

[5.5] TIA/EIA/SP-3693-1 (to be published as TIA/EIA-95-B). *Mobile Station – Base Station Compatibility Standard for Dual-Mode Wideband Spread Spectrum Cellular System*, July 17,1998.

[5.6] TIA/EIA/IS-637. *Short Message Services for Wideband Spread Spectrum Cellular Systems.*

CHAPTER 6

Handoff

The CDMA system supports several types of handoffs, and major categories include hard, soft, and softer handoff. Procedures and parameters for handoff are tailored to maintain call integrity while enhancing CDMA operation.

Hard handoff is a discrete *event* in time when call support is switched from one cell to another and/or from one carrier to another. Soft handoff is a *state* in which the call is simultaneously supported by multiple base stations. The hard handoff event is necessarily brief; in contrast, it is not unusual for a mobile to be in a soft handoff state for a considerable fraction of its call time. Hard handoffs are found in all access technologies (for example, AMPS, TDMA, GSM, CDMA), whereas soft handoff is unique to CDMA.

The following sections further define and discuss handoff types and they describe considerations relevant to parameter settings for soft and softer handoff in some detail and summarize hard handoff strategies for use in multi-carrier CDMA markets.

6.1 Hard Handoff

A brief interruption of the communication link is a characteristic of hard handoff. Examples of hard handoff include handoff from one CDMA carrier to another and a change to a different frame offset. The link must be interrupted if the mobile station switches from one CDMA carrier to another (inter-carrier handoff). In a frame offset change, the link must be interrupted while the mobile station changes the offset of its frame transmissions with respect to system time. In this case, the mobile station remains on the same CDMA carrier.

Other types of same-carrier CDMA hard handoff can occur. For example, in early deployment of CDMA, soft handoff may not be supported between cells that belong to different mobile switching centers (MSCs). Accordingly, same-carrier CDMA-to-CDMA hard handoff will occur across cells on the MSC boundary. The absence of inter-MSC soft handoff can potentially impact system performance along the MSC boundary.

In markets that employ multiple CDMA carriers, hard handoffs between carriers can be quite common. The strategies and procedures for ensuring robust inter-carrier handoffs are essentially extensions of those employed for soft handoffs (see Section 6.2.) but tailored to address the additional scenarios that multiple carriers can create. For example, additional carriers might be employed only for traffic relief in *hot spot* areas, thereby requiring that the mobiles exiting these areas hand off to a ubiquitous (common) carrier before proceeding further. Section 6.3 discusses the parameters and strategies specific to hard handoff for multiple carriers.

6.2 Soft and Softer Handoff

6.2.1 Definition

In soft handoff, multiple cells simultaneously support the mobile station's call. In softer handoff, multiple sectors of the same cell simultaneously support the mobile station's call. The mobile station continuously scans for the pilot signals that each cell/sector (site) transmits and establishes communication with any site (up to three) whose pilot power exceeds a given threshold. Since every cell reuses the same 1.25 MHz channel, these types of handoff do not require an interruption of the communication link.

6.2.2 Procedure

The soft and softer handoff procedures dictate the way to maintain a call as a mobile station crosses boundaries between CDMA cells.

In soft handoff, multiple cells simultaneously support the mobile station's call. In softer handoff, multiple sectors of the same cell simultaneously support the call. Each sector transmits a pilot signal of sufficient power to be detected by mobile stations within its vicinity. The mobile station continuously scans for pilots and establishes communication with any sector (up to three) whose pilot exceeds a given threshold.

Similarly, the mobile station terminates communication with sectors whose pilot drops below a threshold.

Each pilot exhibits a different time offset within the same PN code, and the mobile station's identification of distinct pilot signals relies on this property. These offsets are in integer multiples of a known time delay, facilitating the mobile station's search for pilots. The mobile station categorizes pilots it identifies as well as other pilots that the serving sector(s) specify, as follows.

- The *active set* consists of those pilots whose sites are currently supporting the mobile station's call.
- The *candidate set* consists of those pilots whose sites, based on the received strength of their pilots, could also support the mobile station's call.
- The *neighbor set* consists of those pilots whose sites are not in the active set or the candidate set but are nevertheless likely candidates for soft handoff (for example, these sites may be in known geographic proximity).
- The *remaining set* consists of those pilots within the CDMA system but not within the other three sets.

The mobile station's assessment of pilot signal strength and a set of (adjustable) thresholds determines the movement of pilots among the sets. This movement is coordinated with the serving sector. The mobile station assesses pilots by comparing pilot strengths to one another and by comparing each pilot's power to the total received forward-link power. The latter comparison (normalized pilot strength) is the ratio of the pilot energy in a PN chip time to the spectral density of total received forward-link power:

$$\left(\frac{E_c}{I_o}\right)_i = \frac{\mu P_i / W}{F N_{th} + \sum_{all\ j} P_j / W}, \qquad (6.1)$$

where
- E_c = chip energy received from i-th sector,
- I_o = spectral density of total received interference,
- μ = fraction of sector power allocated to pilot signal,
- P_i = received power from i-th sector,
- W = system bandwidth (1.25 MHz),
- F = base station noise figure,
- N_{th} = thermal noise power density.

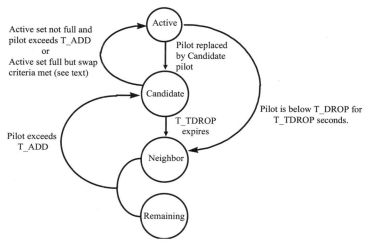

Figure 6–1 Simplified Pilot Set Transactions
Diagram does not show all possible transitions.

Mobile stations associate pilots in the neighbor and/or remaining set whose E_c/I_o exceeds a threshold T_ADD with sites that can support the call; accordingly, these pilots are moved to the active or candidate set. Similarly, pilots in the active and/or candidate set whose E_c/I_o drops below a threshold T_DROP for a period of time exceeding the parameter T_TDROP, are moved to the neighbor or remaining set.[1] Finally, a candidate set pilot whose strength exceeds an active set pilot by at least T_COMP will be moved to the active set, possibly displacing that pilot (swap criteria in Figure 6–1).

Figure 6–1 is a simplified diagram showing the movement of pilots between sets. Rather than attempting to show every possible event, we focus the diagram on those events most influenced by the translatable handoff parameters.

6.2.3 Comparisons

Contrasting this process with the conventional handoff process used in an analog system gives further insight into soft handoff operation. In an analog system, each cell is assigned a set of narrowband channels for use in communication links. Co-channel interference is controlled by not reusing the same channels in adjacent cells. A mobile

---

1. More precisely, the actual threshold in dB equals $-T\_ADD/2$ for the pilot addition and $-T\_DROP/2$ for dropping the pilot since the standard specifies that T_ADD and T_DROP are, respectively, unsigned six-bit binary numbers. For example, when T_ADD (or T_DROP) is set to 11100 (28 in decimal), the pilot add (or drop) threshold is -14 dB. For convenience, T_ADD and T_DROP will be referred to as the threshold in dB throughout this book.

station proceeding out of one cell into another must switch to an available channel in the new cell. This switch requires a brief interruption of the communication link. A CDMA system reuses the same wideband channel in every cell. Co-channel interference is accepted but controlled to achieve greater capacity. Accordingly, soft/softer handoff does not require channel switching and its associated link interruption. Moreover, with proper threshold settings, mobile stations acquire new sites before the old (serving) sites are too far away to be useful. The soft handoff procedure is more robust because the mobile stations connect with the new host(s) before the connection with the old is broken. This process is often referred to as a make-before-break connection as opposed to the analog break-before-make.

The use of a break-before-make handoff procedure (that is, no soft handoff) in a CDMA system could have adverse consequences for system operation. The inability to exploit the diversity gain inherent in multi-site support can cause the areas of cell boundary overlap (the locations furthest from base stations) to be regions of poor link quality. The mobile stations in these fringe areas would also be more susceptible to base station interference. These effects increase the probability that a call will be dropped, since the handoff procedure would typically not be initiated until a mobile station reached this area (that is, until the mobile station noted a drop in signal strength from its host cell). In addition, the use of power control without soft handoff could create a situation where a mobile station generates considerable amounts of interference to neighbor cells. This interference reduces capacity.

The last situation arises because the mobile station would detect a drop in received signal strength before it requested a handoff. Since cell boundaries overlap, this reporting point could be well into the boundary of the neighbor cell. Within this area, power control would boost the mobile station's transmit strength in an attempt to maintain the link with the (distant) serving cell. This call-dragging phenomenon reduces the capacity of the neighbor cell because the mobile station's transmissions increase the level of interference at the neighbor cell. In contrast, if the mobile station were in soft handoff, power control commands from both cells would ensure that the mobile station did not produce undue interference. In fact, the reverse link could be maintained at a lower level of mobile station transmit power due to the gain involved in combining the signals received at the two base stations.

6.2.4 Performance

Judicious use of the soft/softer handoff process can enhance service by raising call quality, improving coverage, and enhancing capacity. In the following section, we discuss the considerations and trade-offs involved in realizing these benefits. Examples of soft handoff performance follow.

6.2.4.1 Considerations

In the following, we consider the impact of the parameter *T_ADD*. Considerations for other parameters (for example, *T_DROP*) are analogous.

The setting of *T_ADD* for each base station will impact site coverage. This value, coupled with the site pilot strengths and area propagation characteristics, will determine those regions in which a mobile station will add the site pilot to its active/candidate set. (Similarly, *T_DROP* will determine regions in which the pilot will be dropped.) These regions, together with typical traffic distributions, will determine the handoff populations, that is, the number of mobile stations in handoff with two or three sites and the number of mobile stations in softer versus soft handoff. he number and location of mobile stations in handoff will impact call quality as well as base station capacity. For example, a parameter setting that allows mobile stations in fringe areas (areas where the link is poor due to distance/propagation) to have soft handoff will improve call quality via diversity gain. Similarly, a setting which places mobile stations in soft handoff before they get too close to neighbor sites will lower the mobile station's required transmit strength, reducing interference and improving capacity. Finally, the setting will impact the probability that a mobile station detects a pilot as well as the probability that it mistakes noise/interference for a pilot.

We can obtain some insight into the above considerations by examining the results of generic simulations using randomly distributed mobile stations and simple path loss laws. We obtained the following results with cells of an eight-mile radius and a 38.4 dB/decade path loss (110.7 dB loss at one-mile intercept). Further methodology for each study is as indicated.

6.2.4.2 Coverage Contour

A coverage contour is the connected set of locations surrounding each cell where the pilot E_c/I_o relative to that cell is equal to the value *T_ADD*:

$$10 \log \left(\frac{E_c}{I_o} \right) = T\_ADD. \tag{6.2}$$

Values of E_c/I_o within the contour will be greater than *T_ADD*; values outside the contour will be less. Accordingly, a mobile station crossing the boundary *into* the cell will add that cell's pilot to its active set. (A mobile station crossing the boundary *out* of the cell will not necessarily drop the pilot as this function depends on the values of *T_DROP* and *T_TDROP*). Coverage areas also change with varying *T_ADD*.

6.2.4.3 Populations

Figures 6–2a and 6–2b show the impact of *T_ADD* on handoff populations for omnidirectional and sectored cells, respectively. A computer model that involves a two-tier service area with randomly distributed mobile stations provided these curves. The model places bases and mobiles within a service area and computes the percentage of mobiles in *n*-cell soft handoff via comparison of the normalized pilot strength measure (6.1) to the threshold *T_ADD*. Averaging over multiple trials incorporated the effects of randomness in mobile placement and in propagation loss.

Some of the following assumptions are used to construct the model.

- The bases are placed in accordance with a nominally hexagonal cell pattern.
- All cells are of equal size and radiate equal power.
- The cell radius, number of tiers, base power, and fraction of base power allocated to the pilot channel are fixed throughout the trials.
- Determination of the signal attenuation from base to mobile location is by spherical spreading out to one mile, and 38.4 dB/decade thereafter; a lognormally distributed random variable multiplies attenuation past the one-mile intercept point.
- The service area consists of all omnidirectional cells or all sectored cells.[2]

The number of mobiles randomly placed into the service area in each trial is chosen to correspond to a fixed, nominal reverse link loading (for example, ninety percent). The number of served mobiles is determined by noting the number of mobiles for which $E_c/I_o \geq T\_ADD$ for at least one pilot. The number of mobiles in a specific handoff state is computed by noting the fraction of served mobiles for which $E_c/I_o \geq T\_ADD$ for the appropriate number of pilots. In each trial, mobiles that do not see at least one pilot at or above *T_ADD* are dropped.

---

2. The sectorization patterns employed were

$$G(\theta) = \begin{cases} 1 - \dfrac{1-b}{(\pi/3)^2}; & |\theta| \leq \sqrt{\dfrac{1-a}{1-b}}\dfrac{\pi}{3} \\ a; & \text{elsewhere} \end{cases},$$

where b = 0.5 and a = 0.0316 (see Reference [6.1]).

More specifically, the following parameters are used in the model:

- cell radius (nominal) = eight miles,
- overall radius of service area = forty miles,
- PN chip rate = 1.2288 Mcps,
- loss at one mile = 110.7 dB,
- standard deviation of lognormal shadow fading = 8 dB,
- base station radiated power = twenty watts,
- base station antenna gain = 10 dB,
- percentage of base station power allocated to pilot channel = twenty percent,
- mobile noise figure = 8 dB.

Figure 6–2a shows the fraction of mobile stations in handoff with n cells (n-cell soft handoff) for omnidirectional cells. Note that, in this nomenclature, one-cell soft handoff is equivalent to no soft handoff. Figure 6–2b shows the fraction of mobile stations in soft and/or softer handoff with n cells (n-cell soft/softer handoff) as well as in softer handoff with a single cell (softer only) for sectored cells. Both curves also indicate the total fraction of mobile stations in a soft handoff and/or softer handoff condition.

The figures indicate how the handoff populations change as the threshold T_ADD is varied. As expected, the total fraction of mobile stations in a soft and/or softer handoff state drops as the threshold gets larger. We see handoff populations for thresholds above $T\_ADD = -11$ dB to be negligible. The two-cell soft handoff curve in Figure 6–2a exhibits a definite maximum across the T_ADD range. As the threshold gets lower, more mobile stations go into handoff and the curve rises; as the threshold is lowered further, more and more of these mobile stations go into three-cell soft handoff and the population in the two-cell soft handoff state must drop. These considerations apply to any n-cell soft or soft/softer curve. In contrast, the fraction of mobile stations in softer-only handoff (Figure 6–2b) remains approximately constant over the T_ADD range. As the threshold is lowered, mobile stations not previously in a handoff state enter into softer handoff with an adjacent sector. At the same time, mobile stations already in this state enter into handoff with a sector from another cell. The curve is flat because the rate of mobile stations entering the softer-only category roughly balances the rate at which they leave.

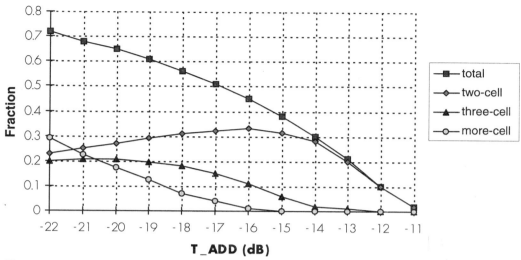

Figure 6–2a Handoff Populations for Omnidirectional Cells

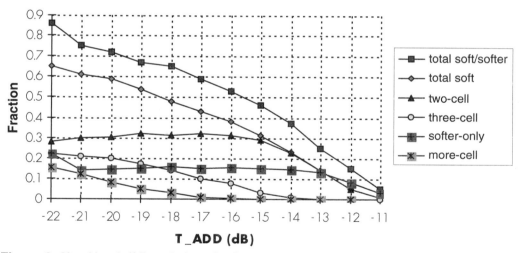

Figure 6–2b Handoff Populations for Sectored Cells

6.2.4.4 *Mobile Station Transmitter Strength*

Reduced mobile station transmit strength in soft/softer handoff is possible because of the gain inherent in the use of multiple sector receivers. Soft handoff realizes this gain at the switch by selecting the best signal (on a frame-to-frame basis) of those

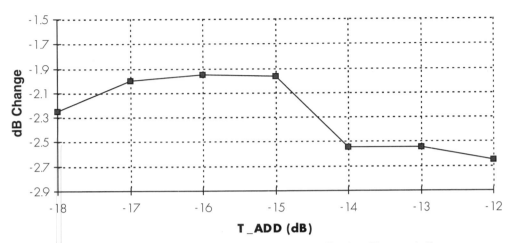

Figure 6–3 Impact of Soft Handoff *T_ADD* on Mobile Station Transmit Power

received. Softer handoff realizes this gain at the base station channel element by combining the signals received from multiple sectors.

Figure 6–3 shows the impact of soft handoff parameter *T_ADD* on mobile station transmit strength. This curve was obtained from a two-tier service area of omnidirectional cells in which mobile stations were randomly distributed. The figure shows the average difference in transmit strength for center cell mobile stations that go into soft handoff. Coherently combining the signals received at the base stations models the diversity gain from multiple base stations.

We plot the average difference in transmit strength as a function of varying handoff threshold *T_ADD*. At high thresholds, only two-cell soft handoff is possible. The average transmit reduction is high because only mobile stations advantageously placed with respect to two cells (for example, mobile stations equidistant between the two) go into soft handoff. At lower thresholds, more mobile stations go into two-cell soft handoff with distant cells: the average reduction is less because one of the signals is slightly weaker and therefore more frequently ignored. The average reduction improves for lower thresholds because some of the mobile stations go into three-cell soft handoff, with commensurately greater diversity gain. The transmitter strengths fluctuate accordingly with *T_ADD* but consistently show a net drop for all thresholds considered.

6.2.4.5 Interference

Section 6.2.4.4 considered the drop in mobile station transmitter strength with varying *T_ADD*. Enabling soft handoff for center cell users in a two-tier service area

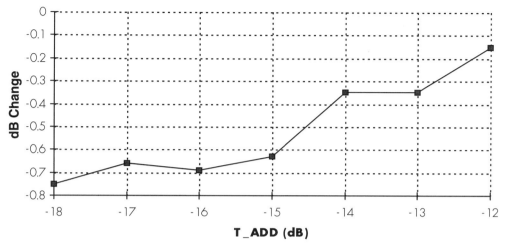

Figure 6–4 Impact of Soft Handoff on Received Power at Center Base Station

resulted in this drop. This section considers the corresponding impact on center base station received power. This impact is of some interest since the level of received base station power is the interference that a mobile station must overcome to establish and/or maintain a call.

Figure 6–4 shows the net impact on the center cell received power. The power shows a general decrease as T_ADD gets lower. The rate of decrease is not constant because it depends on how many mobile stations are in soft handoff and what transmit strength reductions these mobile stations are able to achieve (see Figure 6–3). Since the received power is the interference that a new mobile station must overcome to establish a call, the net decrease indicates that judicious use of soft handoff can improve capacity.

Soft handoff requires the use of a channel element at all base stations supporting the call. The benefit of reduced interference must therefore be balanced against the resource cost of putting a mobile station into soft handoff. Marginal benefit may result if the threshold is low enough to place mobile stations that are close to their serving base stations into soft handoff with distant cells. In this case, the signal frames received at the distant cells would frequently be ignored (that is, not selected) in favor of those arriving at the serving site. Accordingly, little to no interference reduction could accrue. In contrast, selecting handoff thresholds that place mobile stations roughly equidistant between two cells into soft handoff with these cells is fully cost-effective. The switch favors neither received signal and the mobile stations realize the full effect of diversity gain.

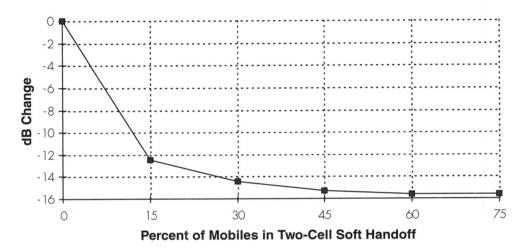

Figure 6–5 Soft Handoff Impact on Center Base Station Received Power (Non-Uniform Mobile Station Distribution)

Examination of the decrease in center base station interference for a non-random distribution of mobile stations in a one-tier service area affords some insight into this comparison. To accentuate the effect, we place the center cell mobile stations with increased density near the center base station and neglect interference from outer cells. The center cell mobile stations are progressively put into soft handoff with the outer cells, proceeding from the outer mobile stations inward. We allow only two-cell soft handoff.

Figure 6–5 shows the drop in base station interference as a function of the percentage of mobile stations in soft handoff. The base station interference drops significantly at first but levels out at around the thirty percent point. Placing further layers into soft handoff beyond this point only marginally lowers the base station interference and is less cost-effective.

6.2.4.6 Search Window Parameters

The parameters discussed in the sections above form a mechanism to control the movement of pilots between sets. They do not, however, provide control over how pilots are actually detected. A companion set of parameters, known as search window parameters, serves this purpose.

Recall that CDMA uses synchronous detection in the forward link. In other words, for the mobile to successfully demodulate any arbitrary pilot signal, it must always have a good estimate of the precise system time. The mobile extracts this estimate from the

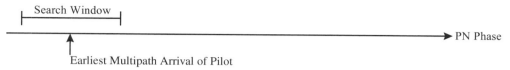

Figure 6-6 Search Window and the Earliest Arrival of the Active Pilot

reference pilot, one of the pilots it is receiving. Given this time reference, the mobile can then synchronously apply any PN code to the signal it receives, thereby extracting the information carried on that pilot.

The catch, however, is that the pilots the mobile is trying to detect will not arrive exactly when expected; the mobile's estimate of system time includes the propagation delay of the reference pilot, but the other pilots will have timing based on their own propagation delays. Since the mobile does not know the amount of propagation delay for any given pilot, it must search over a reasonable window of delays until it finds the pilot's actual timing. How wide a search the mobile makes when looking for a given pilot is called the *search window*, and the suggested settings for these windows and the rules behind them are the topic of this section.

The mobile uses three different search window parameters when searching for pilots:

- *SRCH_WIN_A*, used when searching for active set and candidate set pilots,
- *SRCH_WIN_N*, used when searching for neighbor set pilots,
- *SRCH_WIN_R*, used when searching for remaining set pilots.

6.2.4.6.1 SRCH_WIN_A The mobile station uses this search window setting when it searches for pilots in the active or candidate sets. It is specified in units of PN chips and is centered around the earliest arriving multipath component of the pilot, as shown in Figure 6-6.

The mobile station positions the window so that it searches for arriving multipath components both before and after the current arrival time of the pilot. Thus the mobile will continue to track the pilot in situations of either lessening or growing propagation delay.[3] As a rule of thumb, the search window should be large enough to span the maximum expected arrival time difference between the pilot's usable multipath components

---

3. Lessening propagation delay can occur either gradually or quickly. The former occurs, for example, when the mobile is locked onto a pilot's direct propagation path as it moves toward the cell. The latter occurs, for example, in the case of a mobile being locked on a reflected path as it comes out of an RF shadow into full view of the pilot's direct path (which, by virtue of not being reflected, will arrive at the mobile sooner).

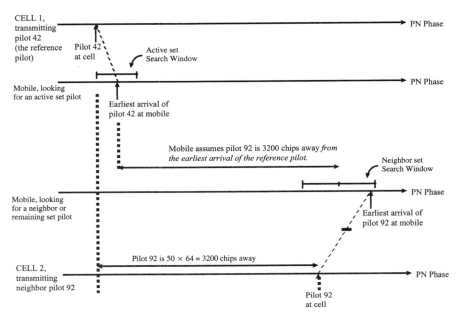

Figure 6–7 Example of Relative Positioning of Neighbor or Remaining Set Search Window

(that is, the pilot's maximum delay spread). An equation to calculate the required delay budget, $T_{d,\text{Active}}$, for this case is

$$T_{d,\text{Active}} > 2 \times \frac{\text{delay\_spread}_{\max}}{T_{\text{chip}}}, \qquad (6.3)$$

where T_{chip} = Chip time (813.8 nanoseconds) and $\text{delay\_spread}_{\max}$ = Maximum delay spread in seconds.

6.2.4.6.2 SRCH_WIN_N and SRCH_WIN_R Mobile stations use these settings when they search for pilots in the neighbor set and remaining set, respectively. Unlike the search window for active set pilots, these windows are centered around the target pilot's PN offset relative to the arrival time of the reference pilot in the active set and this idea is illustrated in Figure 6–7.

For reference, the top two PN phase axes in Figure 6–7 show how the mobile will position the active set search window. The lower two PN phase axes show the relative positioning of the neighbor (or remaining) set search window. The mobile will add the appropriate number of chips to find the neighbor pilot (in this example, finding pilot 92 relative to pilot 42 means adding $(92 - 42) \times 64 = 3200$ chips) based on the position of the earliest arrival of the reference pilot. Since the position of the reference pilot includes its propagation delay, the overall offset to pilot 92 will also include this delay. However, since pilot 92 is coming from a different cell, its propagation delay is different (and, in this example, longer) than that of pilot 42. As a result, the search window size for neighbor and remaining set pilots must account not only for the largest delay spread of the target pilot but also for the largest *difference* in propagation delays (that is, difference in distance) between the reference pilot and the target pilot. An equation to calculate the overall delay budget is

$$T_{d,N} \text{ or } T_{d,R} > 2 \times \frac{D_{max}/v_c + \text{delay\_spread}_{max}}{T_{chip}}, \qquad (6.4)$$

where

D_{max} = Maximum *difference*, in miles, between (1) the mobile and the cell transmitting active set pilot and (2) the mobile and the cell transmitting neighbor (remaining) set pilot,

V_c = Speed of light (186,000 miles/sec),

T_{chip} = Chip time (813.8 nanoseconds),

delay_spread$_{max}$ = Maximum delay spread (seconds).

Having obtained the delay budget for any set, the relationship between the delay budget, the search window size in PN chips, and the value of the search window parameter can be specified, as shown in Table 6–1.

Finally, note that the delay spread, and therefore delay budgets, will depend on the local propagation environment. Urban environments will tend to have more multipath components and therefore more delay spread. Typical values of delay spread for this type of environment is seven microseconds. Suburban environments, on the other hand, have typical delay spread values of around two microseconds. Generally, larger cells will tend to have larger delay spreads than small cells.

6.2.5 Parameters

This section summarizes the use and impact of soft handoff parameters. ANSI J-STD-008 [6.2] specifies these parameters. First there is a brief description of each parameter, followed by a tabulation of recommended parameter ranges. Section 6.2.2 gives the definitions of active, candidate, neighbor, and remaining sets.

Table 6–1 Delay Budget and Search Window Size

Delay Budget (microseconds)	Window Size (PN chips)	SRCH_WIN_A SRCH_WIN_N SRCH_WIN_R
$T_d \leq 1.64$	4	0
$1.64 < T_d \leq 2.45$	6	1
$2.45 < T_d \leq 3.27$	8	2
$3.27 < T_d \leq 4.09$	10	3
$4.09 < T_d \leq 5.72$	14	4
$5.72 < T_d \leq 8.17$	20	5
$8.17 < T_d \leq 11.44$	28	6
$11.44 < T_d \leq 16.34$	40	7
$16.34 < T_d \leq 24.51$	60	8
$24.51 < T_d \leq 32.68$	80	9
$32.68 < T_d \leq 40.85$	100	10
$40.85 < T_d \leq 53.11$	130	11
$53.11 < T_d \leq 65.36$	160	12
$65.36 < T_d \leq 92.32$	226	13
$92.32 < T_d \leq 130.72$	320	14
$130.72 < T_d \leq 184.42$	452	15

The *T_ADD* parameter controls movement of pilots from the neighbor/remaining sets to the active/candidate sets. Mobile stations move a neighbor or remaining set pilot with strength E_c/I_o exceeding *T_ADD* to either the candidate or active set (the decision is based on serving site direction). This parameter is set per sector.

The *T_DROP* and *T_TDROP* parameters control movement of pilots out of the active/candidate sets. The mobile station starts a timer when the strength E_c/I_o of an active or candidate set pilot falls below *T_DROP*. The mobile station moves an active set pilot that falls below *T_DROP* for a period exceeding *T_TDROP* to either the candidate or neighbor set (the decision is based on serving site direction). The mobile station moves a candidate set pilot that falls below *T_DROP* for a period exceeding *T_TDROP* to the neighbor set. *T_DROP* is measured in dB (see Table 6–2); *T_TDROP* is measured in units that map into seconds (see Table 6–3). These parameters are set per sector.

The *T_COMP* parameter controls movement of pilots from the candidate set to the active set. The mobile stations moves a candidate set pilot with strength E_c/I_o exceeding that of an active set pilot by *T_COMP* × 0.5 dB to the active set, replacing that pilot. In situations of four (or more) strong pilots, a low value of *T_COMP* will make it easier for pilots to enter the active set ensuring inclusion of the strongest pilots in the area in the active set. However, this also leads to more pilot interchange activity. Raising *T_COMP* will make it more difficult for pilots to enter the active set, but there is some risk of neglecting a strong pilot that should be included. (Recall that a strong pilot that is not in the active set is a strong interferer.) *T_COMP* is measured in units of 0.5 dB. This parameter is set per sector.

The *NGHBR_MAX_AGE* parameter controls movement of pilots out of the neighbor set. The mobile station maintains an AGE counter for each pilot in the neighbor set and updates this counter under direction of the serving site. The mobile station moves a pilot with AGE count exceeding *NGHBR_MAX_AGE* to the remaining set. This parameter is set per sector.

The *SRCH_WIN_A*, *SRCH_WIN_N*, and *SRCH_WIN_R* parameters govern the mobile station's search for pilots in the active/candidate, neighbor, and remaining sets, respectively. These parameters specify the size of the search window that detects the pilot. The search window for pilots in the active and candidate sets (*SRCH_WIN_A*) is centered around the earliest arriving multipath component. The search window for pilots in neighbor (*SRCH_WIN_N*) and remaining (*SRCH_WIN_R*) sets is centered around the pilot's PN sequence offset. Table 6–4 shows the range and recommended values for the search window parameters. Some important considerations when setting search windows are described below.

- There is a trade-off between search window sizes and search speed. Larger window sizes require more mobile processing per search and thus reduce the overall number of pilots that can be searched in a fixed period of time.
- The mobile will not detect pilots that lie outside the search window regardless of their strength. Thus, a non-detected pilot could become a strong interferer.
- A certain implementation does not admit pilots to the active set if they are not on the neighbor list. In this case, Table 6–4 suggests that, after optimization,[4]

---

4. During optimization, *SRCH_WIN_R* should be set between 7 and 13. This allows the mobile to find pilots that may have been inadvertently omitted from the neighbor list.

Table 6–2 Handoff Parameters

Parameter	Recommended Range	Impact
T_ADD	–13 to –17 dB	impacts site coverage, pilot detection, capacity, handoff populations
T_DROP	–13 to –17 dB	impacts site coverage, capacity, handoff populations

Table 6–3 *T_TDROP* Conversions

T_TDROP	Timer Expiration (seconds)	T_TDROP	Timer Expiration (seconds)
0	<0.1	8	27
1	1	9	39
2	2	10	55
3	4	11	79
4	6	12	112
5	9	13	159
6	13	14	225
7	16	15	319

Table 6–4 Search Window Parameters
(Search window parameter values are given in units that map into window size in PN chips. For the size of search window in PN chips, see Table 6–1.)

Parameter	Range	Recommended Range
SRCH_WIN_A (active/candidate)	0 – 15	5 – 7
SRCH_WIN_N (neighbor)	0 – 15	7 – 13
SRCH_WIN_R (remaining)	0 – 15	7 – 13 during optimization 0 after optimization

SRCH_WIN_R be set to zero. This prevents the mobile from wasting time searching for pilots that cannot be used for handoff.

6.3 Inter-Carrier Handoffs

In CDMA, inter-carrier handoffs are hard, that is, the link is briefly interrupted while mobile and base stations switch from one carrier to another. CDMA supports two basic types of inter-carrier handoff. *Hand-down* is defined as a hard handoff between two different carriers within the same cell. *Handover* is defined as a hard handoff between two different carriers in two different cells, that is, the handoff occurs from one carrier in one cell (source) to a different carrier in another cell (target).

The geometry or area classification imposed by multi-carrier use partly drives the type of inter-carrier handoff employed. Such classifications also provide a convenient means of outlining the various handoff methods. This discussion will consider only two carriers. Multi-carrier areas are, therefore, classified as pocketed (first carrier everywhere, second carrier here and there), disjoint (different regions employ a different, single carrier), or uniform (both carriers everywhere).

The following section considers only pocketed and disjoint areas. Strictly speaking, uniform areas do not require inter-carrier handoffs since the availability of both carriers everywhere permits the independent operation of each carrier (that is, no boundary necessitating an inter-carrier handoff would be encountered). Nevertheless, by using strategies that are straightforward extensions of those described below, inter-carrier handoffs may still be employed in this situation.

6.3.1 Pocketed System

In a pocketed system, the CDMA carrier exists only in isolated, non-contiguous areas (see Figure 6–8). For example, this geometry could occur if a second carrier is added to provide local traffic relief in scattered hot spots throughout the market. Mobiles on the second carrier that exit this pocket of second-carrier cells must hand off to the first (common) carrier to maintain the call.

The most robust way of effecting this handoff is to direct the mobile to hand down to the first carrier *before* it exits the pocket. Handoff across the pocket border then becomes normal soft handoff on the common carrier. Typically, hand-down occurs at the edge of the pocket, that is, within the border cells that separate the pocket from the surrounding (first carrier only) region (see Figure 6–8). In the border cell, the mobile hands down from the second carrier to the first.

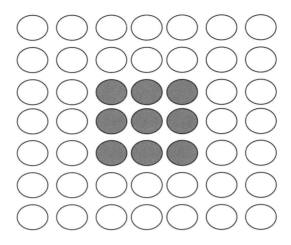

Figure 6–8 Simple Schematic of Pocketed Region

To engineer this process, the border cells (that is, the cells in which hand-down from the second carrier to the first is permitted) must be identified. The identification of a cell as a border cell is on a per-sector basis. In general, a border cell (sector) is any second carrier cell through which the mobile can pass just prior to entering a region with first carrier only. These cells are usually those on the edge of the pocket; however, additional cells further in from the edge should be considered as border if traffic can pass directly through these cells into the first carrier-only region. Careful examination of cell geometry and local traffic routes can greatly facilitate border cell selection.

The base station instructs a mobile entering a sector configured as border to issue frequent, periodic pilot strength measurement messages (PSMMs). This process allows the sector to closely monitor the mobile's situation without waiting for reports triggered by pilot events (see Section 6.2.2). When the pilot report indicates that the sector's (normalized) pilot has dropped below a threshold, the base station directs the mobile to hand down to the first carrier. The threshold used is designed to compel hand-down before the mobile reaches sector edge, thereby allowing enough time and distance for normal soft handoff to occur. It is typically set to a value much larger than *T_DROP*. The hand-down to the first carrier occurs without knowledge of the first carrier pilot strength; however, no ambiguity or compromised coverage should exist since the first-carrier and second-carrier sectors are collocated.

The process described works well for a large pocket in which border cells are easily defined and traffic needs tend to taper off toward the pocket edge. The latter consideration is important because the first carrier in the border cells must have enough capacity to serve both conventional first-carrier traffic (that is, first-carrier originations, terminations, and soft handoffs) as well as the additional traffic imposed by hand-downs from the second carrier.

The identification of border cells is not as straightforward for the smaller pockets that may be created by addressing highly localized traffic hot spots. In these cases, none of the few cells employing a second carrier may have enough excess first-carrier capacity to accommodate hand-downs. This is particularly true for a pocket consisting of a single isolated cell with a second carrier. This geometry presumably exists due to the need to provide first-carrier capacity relief on this single cell. However, the need to hand down from the second carrier to the first before exiting this cell could place a significant amount of second-carrier traffic right back onto the first carrier and thereby recreate the original overload condition.

In situations where the border cell(s) may not have enough first-carrier capacity to accommodate hand-downs, the pocket must be redesigned to shift hand-down operations to cells where sufficient first-carrier capacity is available. Implementing the second carrier in additional cells around the pocket edge accomplishes this redesign; by expansion of the second-carrier pocket so that the border cells become cells that *by design* have sufficient first-carrier capacity to handle normal first-carrier traffic as well as hand-downs.

Cells in which a second carrier is added to facilitate hand-down rather than to provide traffic relief are termed *transition cells*. The area comprising the transition cells is called the *transition zone*. In the case of the single-cell, second-carrier pocket described above, several surrounding transition cells may have to be added to accommodate traffic exiting in any direction. By declaring only the transition cells as border cells, the hand-down from the second to the first carrier occurs within the transition zone rather than within the densely loaded area of the central cell (see Figure 6–8).

Local cell geometry, traffic routes, and cost considerations govern the selection of transition zone cells. In general, a transition zone cell should be added (that is, the pocket should be expanded) whenever the additional inter-carrier handoff traffic imposed upon a border cell could overload the first (common) carrier.

Pocket expansion need not be uniform in all directions; for example, traffic routes and/or natural boundaries might preclude traffic from exiting the pocket in certain directions. In addition, the cost of locally adding transition cell(s) might be weighed

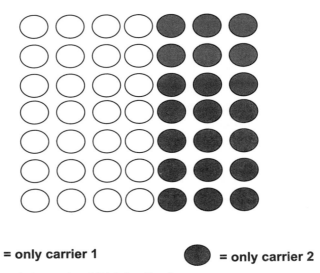

Figure 6–9 Simple Schematic of Disjoint Region

against the cost/risk of call drop in exit areas that are likely to serve only a small amount of second-carrier traffic.

Handover facilitated by pilot-only (beacon) cells may be employed as an alternative method to hand-down in pocketed areas. Section 6.3.2 describes this procedure more fully, with respect to its use in disjoint systems.

6.3.2 Disjoint System

In a disjoint system, distinct CDMA carriers exist within distinct regions (see Figure 6–9), that is, one region has only the first CDMA carrier and the other region has only the second CDMA carrier. This geometry is more unusual but could exist due to regional differences in available spectrum or as a means of facilitating handoffs across a border in special circumstances when no soft handoff is available.

For the purpose of this discussion, the region employing only carrier one will be called *zone one* and the (contiguous) region employing only carrier two will be called *zone two*. In all cases, mobiles crossing the border from zone one into zone two (or vice versa) must hand off between carriers to maintain the call.

Two general methods can effect such a handoff. The first (hand-down) entails creating a border area of cells that support *both* carriers one and two. A mobile entering into the border area hands down from its current carrier to the other carrier before proceeding further. The second (handover) executes the handoff from one carrier to the other as the mobile crosses the border separating the two zones. The handover is not as robust as the hand-down but may be appropriate in certain geometry.

6.3.2.1 Hand-Down

To engineer a hand-down, a border area (that is, the area that supports both carriers and within which hand-down will occur) must be identified and properly configured. In general, the border area should contain all cells that the mobile could pass through just *before* leaving one zone and entering the next. The identification of cells within this area as border cells is on a per-sector basis. The border area is nominally a thin layer (one cell thick) of cells centered along the geometric boundary separating zone one from zone two and thus typically consists of those cells along the edge of the zone(s). Additional cells further in from the edge should also be considered as border if traffic can pass directly through these cells and enter the other zone. Careful examination of cell geometry and local traffic routes can greatly facilitate border cell selection.

The base station instructs a mobile entering a sector configured as border to issue frequent, periodic pilot strength measurement messages. This process allows the sector to closely monitor the mobile's situation without waiting for reports triggered by pilot events (see Section 6.2.2). When the pilot report indicates that the sector's (normalized) pilot has dropped below a threshold (mostly the same as that for the pocketed system hand-down described in Section 6.3.1), the base station directs the mobile to hand down from one carrier to the other. The threshold used is designed to compel hand-down before the mobile reaches sector edge, thereby allowing enough time and distance for normal soft handoff to occur. The hand-down from one carrier to the other is done without knowledge of the other (last) carrier pilot strength; however, no ambiguity or compromised coverage should exist since the first-carrier and second-carrier sectors within the border area are collocated.

The process described works well for a border area where the flow of traffic is linear and predictable, that is, where a mobile crossing from zone one to zone two (or vice versa) is likely to continue into the latter zone. A more complex border area must be created where there is a reasonable probability that the mobile would turn around within the border area and reenter the zone it has just left. A mobile that reverses course within the border sector would not be allowed to hand down to its old carrier again because multiple inter-carrier handoffs within a border sector are not allowed. (This restriction avoids thrashing between the two carriers.) Expansion of the border area width addresses this situation.

Figure 6–10 conceptually depicts this situation. A mobile linearly crossing from zone one (the left side) to zone two (the right side) would hand down from f1 (the first

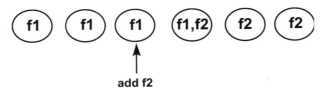

Figure 6-10 Implementation of Double Border

carrier) to f2 (the second carrier) within the single border cell that contains both carriers. If within that cell the mobile reversed course and headed back into zone one, the necessary hand-down from f2 back to f1 would be disallowed: a border cell allows only one inter-carrier handoff.

Expansion of border area width addresses this scenario by implementation of a double border as shown. With a double border, a mobile moving from the left side (zone one) to the right side (zone two) would hand down to f2 in the second (right-most) border cell. After hand-down, if the mobile turned around and reentered zone one, it would hand down back to f1 in the first (left-most) border cell. Analogous considerations would apply for a mobile that originally moves from zone two to zone one and then turns around. The procedure described renders the handoffs robust to direction changes without requiring that a mobile within a border cell execute more than one inter-carrier handoff.

6.3.2.2 *Handover*

In certain situations, handover rather than hand-down may be used to effect inter-carrier handoffs across borders. Although not as robust, this process does eliminate the need to create a border area that deploys both carriers. In handover, a mobile crossing from zone one to zone two simply hands off to the second carrier directly. This process is triggered when the f1 pilot falls below a threshold. Requiring periodic pilot strength measurement messages from the mobile facilitates this process.

To engineer handover requires proper identification and configuration of border cells (that is, cells that will allow handover). In general, the border cells should consist of all cells that the mobile could pass through just *before* leaving one zone and entering the next. The identification of cells within this area as border cells is on a per-sector basis. The border cells are nominally those along the geometric boundary separating zone one from zone two and thus typically consist of those cells along the edge of the zone(s). Additional cells further in from the edge should also be considered as border if traffic can pass directly through these cells and enter the other zone. Careful examination of cell geometry and local traffic routes can greatly facilitate border cell selection.

The base station instructs a mobile entering a sector configured as a border sector to issue frequent, periodic pilot strength measurement messages. This process allows the sector to closely monitor the mobile's situation without waiting for reports triggered by pilot events (see Section 6.2.2). When the pilot report indicates that the sector's (normalized) pilot has dropped below a threshold (typically lower than that for hand-down of 6.3.2.1), the base station directs the mobile to hand over from the carrier f1 in its host cell (in zone one) to the carrier f2 in its target cell (in zone two). The threshold used facilitates hand-over in the vicinity of the border between zone one and zone two (that is, near the cell edge).

In general, handovers are not as robust as hand-downs for effecting inter-carrier handoffs within disjoint areas; however, the handover process may be useful in certain situations. These include boundaries where the implementation of a border area employing both carriers is not practical and/or situations where the anticipated traffic crossing a border is so light that a less robust handoff method can be tolerated.

The robustness of the handover process can be somewhat enhanced by employing beacon (pilot-only) cells just outside each zone. These cells provide pilots on the same carrier employed by the zone they face; for example, pilot-only cells broadcasting an f1 pilot would be placed just outside the f1 zone. In this scenario, a mobile exiting zone one would hand over to f2 in a zone two cell when the beacon's f1 pilot exceeded its host pilot by a pre-specified margin (controlled by the handoff parameter T_COMP). In most cases, the f1 beacon would be collocated with the f2 cell targeted for the handover to eliminate any uncertainty regarding whether the mobile was within the target (f2) cell's coverage area.

6.4 References

[6.1]   S.-W. Wang and I. Wang. Effects of Soft Handoff, Frequency Reuse, and Non-ideal Antenna Sectorization on CDMA System Capacity, *Proc. IEEE Vehicular Technology Conf.*, Secaucus, NJ, pp. 850-854, May 1993.

[6.2]   ANSI J-STD-008. *Mobile Station - Base Station Compatibility Requirements for 1.8 and 2.0 GHz CDMA PCS*, March 1995.

CHAPTER 7

Link Budgets

Network planning and design uses link budgets. In this chapter, we explain the construction of link budgets for an IS-95 system and present two sample link budgets, one for the reverse link and one for the forward link.

Section 7.1 presents the reverse-link budget in an industry-wide standard format and explains each of its entries in detail. In contrast, there is no standardized format for the forward-link budget. Section 7.2 presents an example of a forward-link budget.

7.1 Derivation of the Reverse-Link Budget

In simple terms, a link budget is an accounting of the various losses and gains of a communication link. Consider the IS-95 reverse link shown in Figure 7–1. Let the maximum mobile transmission power be P_m dBm. The mobile antenna has a specific directivity pattern, and let its gain in the direction of the base station be G_m. There is some loss in coupling the energy from the mobile power amplifier to the mobile antenna due to coupler, combiner, and connector loss. Some of the gain is also at times not realized, as the signal from the mobile to the base station may be obstructed by the cell phone user (not a very comforting thought).

The above two losses are combined together, and they are considered as transmit connector, coupler, combiner, and body loss, denoted as L_m. Thus, the maximum transmitted mobile effective isotropic radiated power (EIRP), P_t, in the direction of the cell site is

$$P_t = P_m - L_m + G_m \tag{7.1}$$

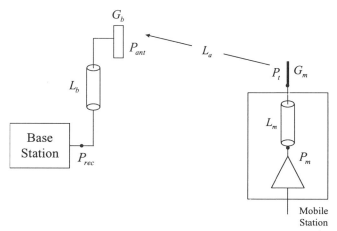

Figure 7–1 IS-95 Reverse-Link

in dBm. This gives us the amount of effective power that the mobile radiates in the direction of the base station.

What is the maximum over-the-air loss (defined in Figure 7–1 as L_a) that this link can tolerate? To determine that, we need to know the signal strength necessary, at the cell site, to maintain a link (or a voice call) at an adequate call quality. One way to measure the call quality is, for example, in terms of a desired frame error rate (FER). Next we evaluate exactly that, the received signal strength necessary at the input to the antenna to maintain a call at the desired reverse FER, denoted as P_{ant}.

7.1.1 Required Power at the Input of the Base Station Antenna

Denote the signal level at the input to the base station receiver antenna from a particular mobile as P_{ant}. Under perfect power control, the base station will receive reverse-link signals from all the mobiles that are power controlled by this particular sector at the same power level. Let the gain of the base station antenna in the direction of the mobile be G_b (where the subscript "b" is used to denote all base-station-related quantities, just as "m" was used to denote all mobile-related quantities), and the loss in the cable connecting the antenna to the base station be L_b. The power level at the input to the cell-site receiver, denoted as P_{rec}, can be calculated as

$$P_{rec} = P_{ant} + G_b - L_b. \tag{7.2}$$

Next we attempt to calculate P_{rec}, the power required at the input of the CDMA receiver to maintain a desired FER. This term is also referred to in the link budget and in the

Derivation of the Reverse-Link Budget

CDMA literature (including the IS-95 standards) as *receiver sensitivity*. The dimensionless quantity E_b/I_o is a measure of the quality of a communication link in a CDMA (more generally, for any digital) system. Next, we evaluate E_b/I_o in terms of the received power, P_{rec}.

The energy-per-bit at the receiver input is given as

$$E_b = \frac{P_{rec}}{R} \text{ (Joules/bit)}, \quad (7.3)$$

where R denotes the data rate of the system. For a vocoder that operates at 13 Kbps, the system data rate is 14.4 Kbps, and for an 8 Kbps vocoder, the data rate is 9.6 Kbps. As we go through the derivation, we will keep a check on the dimensions of each of the terms. Such a dimensional analysis in this case helps in a better understanding of the derivation. P_{rec} is in dBm, which is equivalent to Joules/sec. (More precisely, $10^{P_{rec}/10}$ is in Joules/sec or Watts.) Because R has dimensions of bits/sec, E_b has the dimensions of Joules/bit, which indicates why E_b is referred to as energy-per-bit.

Next we calculate the amount of interference (thermal noise, receiver noise, and CDMA signals from other users) that the reverse-link signal has to combat. Thermal noise floor density (expressed in terms of energy per Hz) at the input of the cell-site receiver is given as KTF, where K is Boltzmann's constant (equal to 1.38×10^{-23} Joules/Kelvin) and T is the temperature in degrees Kelvin (typically taken as 290 °K). Thus[1] KT works out to be -174 dBm $\times$ seconds (or dBm/Hz), and F is the receiver noise figure (a typical value is ≈ 3.2 (or 5 dB)).[2] This is the minimum noise floor that any communication system has to combat.

Any additional non-CDMA in-band interference at the cell site adds to the above thermal and receiver noise and raises the noise floor. Such an interferer would essentially reduce the maximum allowable path loss, which raises the received power necessary at the base station to maintain signal quality, and hence it results in shrinkage of the cell size.

In addition to the above noise, in a CDMA system, signals from other CDMA users (on the same sector, as well as from other sectors) act as thermal noise *like* inter-

---

1. $KT = 1.38 \times 10^{-23} \times 290$ Joules (or Watts $\times$ seconds). As 1 Watt = +30 dBm, converting KT to dBm, we have $KT = 10 \times \log(1.38 \times 290) + 10 \times \log(10^{-23}) + 30$ dBm $\times$ seconds.
2. For more details on how a system noise figure is calculated and referenced at the input of a micro/RF receiver for the purpose of *signal to noise type calculations*, refer to any standard text on microwave receivers (for example, [7.1]).

ference, which raises the noise floor by the receiver interference margin.[3] Thus, the total interfering noise, consisting of CDMA internal interference, receiver noise, and thermal noise, is given as

$$I_o = KTF \times R_{im}, \qquad (7.4)$$

where R_{im} is the receiver interference margin and is taken as 2 (that is, 3 dB) in the example link budget presented at the end of this section. As F and R_{im} are dimensionless numbers, the dimension of I_o is dBm × seconds (or dBm/Hz), which is the same as Joules.

Thus the dimensionless quantity, the ratio of energy-per-bit to interference density (denoted as E_b/I_o), can be obtained from (7.3) and (7.4) as

$$\frac{E_b}{I_o} = \frac{P_{rec}}{R \times KTF \times R_{im}}. \qquad (7.5)$$

Depending on the E_b/I_o necessary to maintain the quality of the voice call, using (7.5), one can calculate the power required at the receiver input. For the IS-95 reverse link, an E_b/I_o value of 7 dB may be taken as a reasonable average value to maintain a call at the FER of approximately one percent. This number comes from the simulation results shown in Figures 7–2a - 7–2c, which show the E_b/I_o required to maintain one percent FER for various speeds and different numbers of paths. As there is diversity on the receive path due to two receive antennas at the base station, there are at least two paths on the reverse link. Hence it is appropriate to use two-path numbers. The one-path and four-path cases attached give an idea of the diversity gain (for example, comparing the one-path case to the two-path case) and indicate the decreasing marginal advantage of having more than two paths.

For the two-path case, 30 Km/h is the worst-case speed, requiring an E_b/I_o of 7 dB to obtain a FER of one percent. Hence the link budget example uses the E_b/I_o value of 7 dB.

We also see that there is an explicit term for diversity gain in the link budget (line "o" in the example reverse-link budget shown in Table 7–1). By using the E_b/I_o value for the two-path case, we have made implicit use of the diversity gain, and hence we will put a value of 0 dB for *explicit diversity gain*. However, from a comparison of

---

3. See Equation (8.6) in Chapter 8 for an explanation and derivation of the term *receiver interference margin*, including the rise in noise above the thermal noise floor.

Derivation of the Reverse-Link Budget

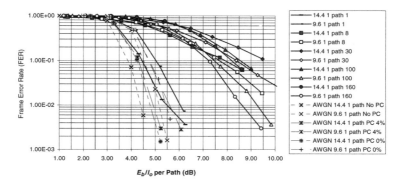

(a) Reverse-Link Performance at 1870 MHz with One Path (for Reference only)

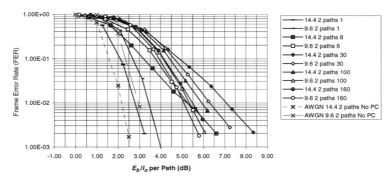

(b) Reverse-Link Performance at 1870 MHz with Two Equal-Average-Power Paths

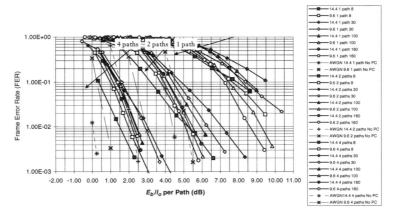

(c) Reverse-Link Performance Summary at 1870 MHz

Figure 7–2 Reverse-Link Simulation Results

Reproduced from [7.2] under written permission of the copyright holder (Qualcomm, Inc.)

Figures 7–2a and 7–2b, we see that the diversity gain is anywhere in the range of about 3 to 6 dB, depending on the mobile speed.

In addition to the above considerations, there is one more aspect that we need to consider. The received signal in a cellular system has large fluctuations. A high-level view of this, which is covered in the next section, is as follows. Because of these fluctuations in the received signal strength, we have to leave additional margin to guarantee link reliability. Related terms in the link budget are fading margin and soft handoff gain. Next we explain these two terms.

7.1.2 Fading Margin

Figure 7–3 shows path loss in dB as a function of distance. What we observe from this path loss data is that, at any given distance, there is a significant variation in the path loss about its mean value in dB scale or its median value in linear scale.[4] At times, a change in receiver position by a few feet results in a change in the signal strength by as much as 10 to 20 dB. This happens as the RF path is obstructed by some man-made structure or natural formation. This phenomenon is called *shadow fading*, or *slow fading*,[5] and it results in the scatter in received signal strength at a fixed distance from the cell site. For example, the path loss at a distance of two miles (that is, at horizontal axis value of 0.3 in Figure 7–3) varies from as much as 147 dB to as low as 130 dB in the data shown in Figure 7–3. As networks are designed using the mean path loss at a certain distance, to guarantee adequate signal strength (and consequently adequate performance), a certain margin has to be left to combat this variation in signal strength. In the link budget, this term is referred to as *fading margin*. We next present a typical model to predict an average value for the path loss versus distance, together with a derivation of the required fading margin.

Let the standard deviation of the variation in signal strength (in dB) at a fixed distance about its mean value be σ. Empirical evidence indicates that the standard deviation is a function of the distance, typically increasing as the distance increases. However, in most cases, a single number represents the shadow fading standard deviation. A typical value for this number for an outdoor (on street) environment is 8 dB.

---

4. When the variation (in dB) is characterized by a normal distribution, the mean value in dB corresponds to the median value in linear scale, that is, if $x = 10 \log y \sim N(m_x, \sigma_x^2)$, then the median of y equals $10^{m_x/10}$.
5. Rayleigh fading or fast fading, in contrast to shadow fading or slow fading, can result in rapid signal strength changes from movements of the order of a wavelength ($\approx$ 6.5 inches at PCS frequency and $\approx$ 13.8 inches at cellular frequency). Rayleigh fading occurs due to the constructive and destructive interference of radio waves. Fast fading is essentially tackled in IS-95 with interleaving and coding. The purpose of reverse-link power control is to combat slow fading and to mitigate the near-far problem.

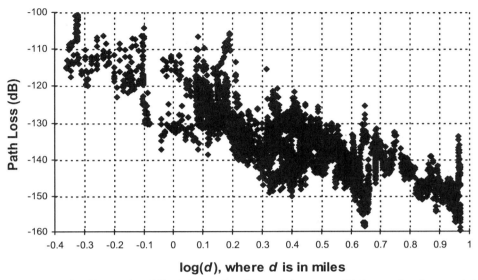

Figure 7–3 An Example of Path Loss (in dB scale) versus Distance (on log scale)

The typical model used to predict signal strength at a given distance is a lognormal distribution (that is, a normal distribution in dB scale). That is, propagation loss at a distance d to a mobile, as measured from a base station, is a random variable modeled as $L(d,\zeta) = d^\gamma 10^{\zeta/10}$, where γ is the path loss exponent (for example, $\gamma = 4$ for a fourth power law) and ζ is the attenuation in dB due to shadowing effects, that is, modeled by a Gaussian random variable with zero mean and a standard deviation of σ, and contributes to fluctuations in the path loss. In dB, the path loss is given as

$$10\log[L(d,\zeta)] = 10\log[d^\gamma 10^{\zeta/10}] = 10\gamma \log d + \zeta . \tag{7.6}$$

Observe from (7.6) that as measured in dB, the path loss at a distance d from the cell site consists of a deterministic component (mean value) of $10\gamma \log d$ and a zero mean Gaussian distributed random component with a standard deviation of σ. We observe that *without shadowing* (the case when $\zeta = 0$), the amount of power required to offset the propagation loss is a deterministic quantity. This is given by $10\gamma \log d$. We also observe that the handoff of a call will occur in this ideal case when a mobile crosses over the edge of one hexagonal cell to another or when the minimum distance switches from one base station to another.

How much additional margin is needed to account for shadow fading? Before we answer this question, let us consider one of the main uses of link budgets. Link budgets

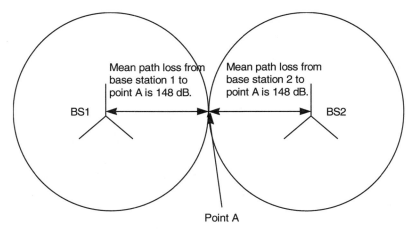

Figure 7–4 Mean Path Loss from Base Station to Cell Boundary

are used for network planning. They are used to ask the question, How far apart can we place two base stations so that there is adequate coverage between the two cells? Suppose that we place two cells as shown in Figure 7–4 so that, at the edge of the two cells, the median (or mean[6]) path loss is 148 dB from either of the two cells. Due to the shadow fading phenomenon discussed above, the path loss at points along the boundary (such as point A) will be greater than 148 dB fifty percent of the time and will be less than 148 dB fifty percent of the time. Also, at many points inside the cells, the minimum (to either of the two cells) path loss may exceed 148 dB. Let us define outage as the case when the minimum path loss is greater than the median value (148 dB in this case). If no additional margin were left, then fifty percent of the points at the edge would be in outage. Such a high probability of outage would not be permissible in many designs, and hence the link budget includes additional margin to reduce the probability of outage or to increase the probability of coverage at the edge of a cell.

Next we calculate the additional margin. In contrast to the non-shadowing case considered before, consider the case when *shadow fading is present*. Suppose that we want the link to have adequate power to compensate for variation in propagation loss for ninety percent of the bins on the edge of a cell. In that case, apart from the $10\gamma \log d$ dB, we need an additional link margin ρ such that the shadowing random variable ζ takes on a value that is less than ρ ninety percent of the time. Its complementary

---

6. When the path loss in dB is modeled by a Gaussian random variable, its mean and median are the same.

Derivation of the Reverse-Link Budget

event (when the link fails to have an adequate margin) is defined as the outage probability. The outage probability is then given by

$$P_{out} = \Pr(\zeta > \rho) = \frac{1}{\sqrt{2\pi}\sigma} \int_{\rho}^{\infty} e^{-\zeta^2/2\sigma^2} d\zeta = Q(\frac{\rho}{\sigma}), \qquad (7.7)$$

As ζ has a Gaussian distribution in dB, the integral of the Gaussian distribution is expressed in terms of the Q function[7] in (7.7). If the link has to have adequate power ninety percent of the time for points on the cell edge, then $P_{out} = 0.1$, which works out to $\rho/\sigma \approx 1.29$, or $\rho = 1.29\sigma$. For $\sigma = 8$ dB, $\rho = 10.3$ dB. This is the additional margin required to maintain the link at ninety percent of the points along the edge of a cell and would be the fading margin for this specific case. In fact, in addition to the 10.3 dB, Reference [7.3] claims that an additional margin of 2 to 4 dB is required to avoid the so-called *ping-pong* effect at the cell edge.

Below are a few of the values (from the standard tables of the normal distribution) for the extra margin needed to obtain the desired probability of edge coverage:

Probability of outage = 10 percent $\Rightarrow$ Prob. of edge coverage = 90 percent $\Rightarrow Q(\rho/\sigma) = 0.10 \Rightarrow \rho = 1.29\sigma$,
Probability of outage = 25 percent $\Rightarrow$ Prob. of edge coverage = 75 percent $\Rightarrow Q(\rho/\sigma) = 0.25 \Rightarrow \rho = 0.68\sigma$,
Probability of outage = 50 percent $\Rightarrow$ Prob. of edge coverage = 50 percent $\Rightarrow Q(\rho/\sigma) = 0.50 \Rightarrow \rho = 0$ dB.

Network design interprets the above margin in the following way: let us say that 148.1 dB is the maximum median allowable path loss (as in line "q" in the example link budget presented later in Table 7–1). If a network were designed where points at the edge of any two cells have a median path loss of 148.1 dB and that median path loss was obtained after a fading margin of 10.3 dB was left in the link budget, then from a design point of view, such a network would be expected to have ninety percent probability of edge coverage. The above calculation of fading margin is independent of the multiple access technologies and would apply to CDMA, AMPS, TDMA, and GSM systems.

7.1.3 Soft Handoff Gain

In a CDMA system, an advantage due to soft handoff gain results in effectively lowering the margin required to obtain a specific probability of edge coverage, as compared to other technologies. The soft handoff gain calculation methodology sketched out below follows the development in [7.3]. For a CDMA system that admits soft hand-

---

7. $Q(x) = \int_{x}^{\infty} \frac{1}{\sqrt{2\pi}} e^{-\frac{t^2}{2}} dt$.

off, for any given frame, the switching center will utilize the better (or alternatively, stronger) of two or more base stations' reception. For simplicity, consider that the decision will depend only on the attenuation and that the base station with the lesser of the two (or more) attenuations will control the mobile. The attenuation of a mobile to base station i is given by

$$10\log[L(d_i, \zeta_i)] = 10\gamma \log d_i + \zeta_i, \qquad (7.8)$$

where d_i is the distance to base station i and ζ_i is the corresponding lognormal shadowing variable. One problem is that the random component of the attenuation to the different base stations (the various values of ζ_i, $i = 0, 1, 2, ...$) could be correlated with one another. To get around that problem, the various values of ζ_i are alternately expressed in terms of two independent random variables. Following along the same lines as the development in [7.3], we define $\zeta_i = a\xi + b\xi_i$, where, $a^2 + b^2 = 1$ (see, also, (4.18) in Chapter 4). The idea here is that by using different values for a and b, we can vary the correlation between the various values of ζ_i. $a = 1$, $b = 0$ is the completely correlated case, while $a = 0$, $b = 1$ represents the completely uncorrelated case. For numerical calculations, values of $a = 1/\sqrt{2}$, $b = 1/\sqrt{2}$, a partially (fifty percent) correlated case, will be considered.

Next we evaluate the excess link margin required in this case. Consider the scenario when a mobile is tracked by three base stations, or alternately, is in three-way handoff. Figure 7–5 shows this situation. Link outage will occur in this case only if attenuation to all the three base stations is greater than the margin ρ. Hence

$$P_{out} = \Pr\{\min[10\gamma \log d_0 + \zeta_0, 10\gamma \log d_1 + \zeta_1, 10\gamma \log d_2 + \zeta_2] > \rho\}. \qquad (7.9)$$

Even before we evaluate the above expression, just a glance at the equation gives us an idea of why we have gain due to soft handoff. Instead of a single random variable, ζ, being greater than a fixed value resulting in an outage, we now need two (or three) partially independent random variables, each of which has to be greater than the fixed value to have an outage. The probability of the latter event occurring is certainly less than the former or, alternatively, for the same outage probability, we need less margin in the latter case. This is the advantage of the soft handoff capability in reducing the margin required, which effectively translates into soft handoff gain. For a mobile in soft handoff, even if any one of the three links is not faded, that is good enough to maintain a strong link.

We do not go into the details of evaluating (7.9) and refer the interested reader to [7.3]. For $a = 1/\sqrt{2}$, $b = 1/\sqrt{2}$, and $\sigma = 8$ dB, the soft handoff gain numerically works out to be approximately 4 dB. Due to the soft handoff feature, excess link margin

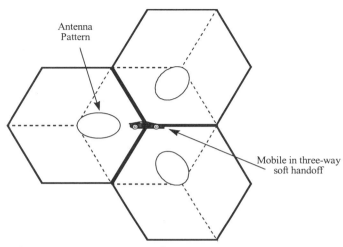

Figure 7-5 A Scenario of Three-Way Soft Handoff

requirement has dropped by 4 dB, from 10.3 dB to 6.3 dB. This is the advantage due to soft handoff that results in increased coverage.

7.1.4 Maximum Supportable Over-the-Air Path Loss

A link budget analysis indicates the maximum supportable over-the-air path loss for a communication system by accounting for the various gains and losses in the path of the communication link. Now we are in a position to calculate how much over-the-air loss the IS-95 link can tolerate.

Denote the soft handoff gain as S_h and the shadow fading margin as S_f. Equations (7.2) and (7.5) give the signal strength required at the input to the base station antenna, and Equation (7.1) states the available EIRP at the output of the mobile antenna. Thus the maximum air loss that can be tolerated, L_a (in dB), is given as

$$\begin{aligned}L_a &= P_t - P_{ant} - (S_f - S_h) = P_t - (P_{rec} - G_b + L_b) - (S_f - S_h) \\ &= P_t - P_{rec} + G_b - L_b - S_f + S_h.\end{aligned} \quad (7.10)$$

Equation (7.10) gives the maximum median over-the-air path loss that can be tolerated between the antenna of the mobile and the base station antenna. Next we present a sample reverse-link budget for a CDMA PCS (1900 MHz) system.

Link budgets are useful in understanding the impact of various hardware changes and performance trade-offs for a system. For example, What is the impact of a 2 dB higher gain antenna or a 3 dB increase in antenna cable loss? How much additional cov-

Table 7–1 Reverse-Link Budget of a 14.4 Kbps CDMA PCS System with a Cell Boundary Coverage Probability of Ninety Percent for Mobile Station Amplifier EIRP of 200 mW

Item	Units		Comments and the Corresponding Notation Used in the Text to Denote the Specific Item
(a) Maximum Transmitted Power per Traffic Channel	dBm	23	As per PCS standards [7.4], a PCS mobile should be capable of 200 mW EIRP: $P_m + G_m$.
(b) Tx Coupler, Connector, Combiner, and Body Loss	dB	2	L_m
(c) Transmitter Antenna Gain	dBi	0	Antenna Gain G_m is assumed as 2 dBi, and it is included in item (a).
(d) Transmitter EIRP per Traffic Channel (a − b + c)	dBm	21	This corresponds to P_t, and the equation used is equation (7.1).
(e) Receiver Antenna Gain	dBi	17.5	G_b
(f) Receiver Cable and Connector Losses	dB	1.5	L_b
(g) Receiver Noise Figure	dB	5	F
(h) Receiver Noise Density	dBm/Hz	−174	KT
(i) Receiver Interference Margin	dB	3.0	R_{im}, see equation (8.6). This noise rise corresponds to twelve users per sector.
(j) Total Effective Noise plus Interference Density (g + h + i)	dBm/Hz	−166	$I_o = KTF \times R_{im}$, equation (7.4)
(k1) Information Rate 10log(R)	dB	41.6	R
(l1) Required E_b/I_o	dB	7	E_b/I_o
(m) Receiver Sensitivity (j + k + l)	dBm	−117.4	Notation used in text is P_{rec}, and the equation used to derive it is (7.5).
(n) Handoff Gain	dB	4	S_h
(o) Explicit Diversity Gain	dB	0	
(p) Lognormal Fade Margin	dB	10.3	S_f
(q) Maximum Path Loss {d − m + (e − f) + o + n − p}	dB	148.1	Equation (7.10)

Derivation of the Reverse-Link Budget

erage (or a reduction in the number of cell sites necessary) can one obtain due to a change in edge coverage reliability from ninety percent to seventy-five percent? What is the reduction in coverage due to an increase in capacity per sector from twelve users per sector (as in this link budget example) to fourteen users per sector? The impact of such changes on the system can be assessed using a link budget.

Observe that the supportable path loss of 148.1 dB is in the direction of the antenna boresight, where the antenna gain is taken as 17.5 dBi (or 15.3 dBd) at boresight. The supportable mean path loss will be correspondingly less in other directions. If the mobile is in a direction where the antenna gain is x dB less than 17 dBi, the maximum allowable path loss in that direction will be x dB less. A cable loss of 3 dB more (for example, if longer cables are required) essentially translates into 3 dB less over-the-air loss that can be tolerated. If maximum antenna gain is increased by 2 dB by using higher gain antennas, the result would be a 2 dB increase in the maximum allowable path loss in the boresight direction. Also, as indicated before, this value of 148.1 dB is the path loss that is exceeded fifty percent of the time and is the median path loss value for which cells should be sized in doing an IS-95 network design. As the link budget considers an additional margin of 6.3 dB (item (p) – (n) in Table 7–1, or $S_f - S_h$), if cells are sized as indicated above, the link will have adequate performance at ninety percent of the bins on the cell edge. Also note that this link budget does not include in-vehicle path loss. The path loss indicated in the link budget, L_a, is the maximum that can be tolerated between the mobile transmit antenna and the cell-site receiver antenna. Sometimes it is referred to as the *street-level* path loss, as in the case when the mobile is typically on the street. If this mobile is inside a building, then an additional penalty for in-building penetration needs to be considered. For in-vehicle coverage, an additional penalty (typically about 6 dB) should be considered. For the example considered above, that means that if a network were to be designed for in-vehicle coverage, then the maximum allowable on-street median path loss at the edge of a cell would be 148.1 – 6 = 142.1 dB.

The receiver interference margin of 3.0 dB (item (i) in Table 7–1) comes by considering a loading of fifty percent of pole capacity, along with a mean voice activity factor of 0.4, and another cells- or sectors-to-in-sector-interference ratio of 0.85. We assume an E_b/I_o value of 7 dB per antenna (a reasonable average value for the reverse link) in the above calculations. For the above set of assumed parameter values, the pole point works out to twenty-four and the number of active voice channels works out to twelve per sector.

All other things remaining the same, suppose we want to support fourteen instead of twelve users per sector in the above example. That would correspond to a fractional loading of 14/24 = 0.5833 (see also Figure 8–2b and Equation (8.6)), and hence a receiver interference margin (R_{im}) of 3.8 dB. Thus the maximum path loss would decrease by 0.8 dB, which would result in a reduction in the coverage of the cell. We take this example further, assuming a path loss slope of 40 dB/decade. Then, we find that the cell radius would shrink by –0.8 = 40log [Fractional Reduction in Radius] => Fractional Reduction in Cell Radius to 0.955, or ninety-five and a half percent of the original.

Thus the trade-off here is that each cell could carry two additional users, but the cells would have to be brought closer so their radius would reduce by about four and a half percent. This is an illustration of how system trade-offs are studied through link budgets.

7.2 Forward Link

This section presents, purely for illustrative purposes, a sample forward-link budget. As stated before, the forward-link air interface does not lend itself as readily to a tabular representation of the link budget as the reverse-link, and hence the example in this section should be considered only illustrative. One of the reasons for this difficulty in tabular representation is the wide variation in the E_b/I_o requirements on the forward link, depending mainly on the speed of the mobile station and the number of paths that the Rake receiver receives [7.5]. Another reason is the unavailability of dedicated power to each traffic channel on the forward link. Instead, all the users share the total available traffic power. We present the link budget next and follow that with a discussion of some of the items.

In this example, we consider twelve users, all of them along the edge of an embedded base station. We assume all of the users to be in two-way soft handoff and all of them see two independent signal paths, that is, all of them are taken as two-path Rayleigh-faded cases. For the above mobile stations, if they are traveling at 26.9 mph or faster, an average E_b/I_o of about 8 dB per path is required to maintain a FER of one percent (see Reference [7.5]). Note that one percent FER is not strictly required and acceptable voice quality may be achieved at higher FER levels. We see that the traffic channel E_b/I_o in line 47 is about 8 dB.

Most of the table is self-explanatory. The derivation of an item from the previous one is given in the comments field. We comment here on some of the independent inputs to Table 7–2. The overhead factor in line 12 is two in this example, as we assume all the mobile stations to be in two-way soft handoff. In an actual system, this overhead factor would be different, and a value of 1.85 is recommended in Reference [7.6]. The mean VAF value taken in line 15 also corresponds to the case of a two-way soft hand-

Table 7–2 Forward-Link Budget of 14.4 Kbps CDMA PCS System with an 8 W Long-Term Average Power Base Station Amplifier

A	B	C	D	E	F	G
Line No.	Description					Comments
	Transmit Power Calculations					
5	Nominal Power Available at Antenna port	8	W	39.0	dBm	Maximum power available
6	Pilot Channel power	1.2	W	30.8	dBm	Set at 15% of max. power
7	Sync Channel Power	0.12	W	20.8	dBm	Set at 10% of pilot power
8	Paging Channel Power	0.42	W	26.2	dBm	Set at 35% of pilot power
9	Power Available for the Traffic Channel	6.26	W	38.0	dBm	78.2% of total power
10	Total Overhead	21.8	percent			C10 = 100*(1 − (C9/C5))
11	Number of Mobiles per Sector	12		10.8	dB	Total number of active users
12	Overhead Factor to Convert from Mobiles to the Number of Active Power Channels	2		3.0	dB	Due to users being in two-way handoff in this case
13	Total Number of Active Power Channels	24		13.8	dB	C13 = C11 * C12; the number of calls supported by the transmitter
14	Average Traffic Channel Power per User	0.26	W	24.2	dBm	C14 = C9/C13; mean power
15	Mean Voice Activity Factor (VAF) for Calls in Two-Way Soft or Softer Handoff	0.479				As all mobiles are in two-way handoff for this example
16	Peak Traffic Channel Power per User	0.54	W	27.4	dBm	C16 = C14/C15; power at full rate of 14.4Kbps
17	Cell Site Cable Loss	1.41		1.5	dB	
18	Cell Site Transmit Antenna Gain	56.23		17.5	dBi	
19	Traffic Channel EIRP per User at Full Rate	21.68	W	43.4	dBm	C19 = C16 * (C18/C17)
20	Total EIRP	318.49	W	55.0	dBm	C20 = C5 * (C18/C17)
21	**Propagation Loss**					
22	Max. Mean Propagation Path Loss	6.46E + 14		148.1	dB	
23	Lognormal Shadow Fade Margin	2.69		4.3	dB	Value obtained off-line
24	Total Allowable Path Loss	1.74E + 15		152.4	dB	
25	**Mobile Receive Signal Power Calculations**					
26	Mobile Receive Antenna Gain	1.58		2.0	dBi	

Table 7–2 Forward-Link Budget of 14.4 Kbps CDMA PCS System with an 8 W Long-Term Average Power Base Station Amplifier (Continued)

27	Mobile Body Loss	1.58		2.0	dB	

A	B	C	D	E	F	G
Line No.	Description					Comments
28	Mobile Receive User Signal Power at Full Rate	1.25E – 14	W	–109.0	dBm	C28 = (C19*C26)/(C24*C27)
29	Mobile Receive Total Power from the Serving Cell	1.83E – 13	W	–97.4	dBm	C29 = (C20*C26)/(C24*C27)
30	**Interference Power Calculations**					
31	Other Users' Orthogonality Factor	0.16		–8.0	dB	From same sector's other Walsh channels
32	Other Users' Interference for E_b/I_o Calculation	2.81E – 14	W	–105.5	dBm	C32 = C31 * (C29 – C28 * C15)
33	Ratio of Mean Other Sector Interference to Same Sector Power at Cell Edge	0.40		–4.0	dB	Value obtained off-line from simulations
34	Other Cells Interference Density	7.30E – 14	W	–101.4	dBm	C34 = C33 * C29
35	**Thermal Noise Calculations**					
36	Mobile Noise Figure (F)	10.23		10.1	dB	
37	Thermal Noise Density ($N_o = KT$)	3.98E – 21		–174.0	dBm/Hz	
38	Total Thermal Noise Power per Hz ($N_o F$)	4.07E – 20		–163.9	dBm/Hz	C38 = C37 + C36
39	Spreading Bandwidth (W)	1.23E + 06	Hz	60.9	dB	
40	Total Thermal Noise Power ($N_o W F$)	5.01E – 14	W	–103.0	dBm	
41	External (intermod/spectrum clearance) Interference	1.56E – 15	W	–118.0	dBm	
42	Total Interference to the Traffic Channel ($I_o W$)	1.53E – 13	W	–98.2	dBm	C42 = C40 + C34 + C32 + C41
43	Total Interference to the Traffic Channel per Hz (I_o)	1.24E – 19	W/Hz	–159.1	dBm/Hz	C43 = C42/C39
44	**Bit Energy to Interference Calculations**					
45	Traffic Channel Bit Rate	14400	bps	41.6	dB	Bit rate for the 13Kbps vocoder
46	Energy per Bit at Full Rate (E_b)	8.66E – 19	W/Hz	–150.6	dBm/Hz	C46 = C28/C45
47	Traffic Channel E_b/I_o	6.97		8.43	dB	C47 = C46/C43

Cell Boundary Coverage Probability of 90 Percent; Pilot Channel Power Allocation 15 Percent; Paging Channel Power Allocation 5.27 Percent; Sync Channel Power Allocation 1.5 Percent; Traffic Channels Power Allocation 78.2 Percent. All power values indicated are long-term average values.

off. The mean path loss is taken as 148.1 dB and the number of active mobile stations is taken as twelve to balance the two links. All of the traffic channels present share power on the forward link. Also, limits on the forward-link traffic channel digital gain setting restrict each individual channel's power allocation. Taking power sharing and the digital gain restrictions into consideration, an additional margin of 4.3 dB (line 23) is needed in this case to ensure link performance ninety percent of the time at the edge. Other users' orthogonality factor (line 31) comes about as a consequence of multipath effects and imperfect implementation of the orthogonality among same-sector Walsh codes. It is taken as 8 dB below the total power from all the other Walsh functions of the same sector. Line 33 is the ratio of the power from other base stations not in soft handoff to the power received from the same base station. For mobile stations at the edge of the base station, we obtain this number from simulations to be about –4 dB. For mobile stations in the interior of a base station, this number would, in general, be lower. Thus, this particular case of a desired E_b/I_o of 8 dB requires an amplifier power of about eight watts at the antenna terminal. However, it should be realized that this is an example of a case where all users are assumed to be identical and that in general the power requirement of each user will be different, depending on the user's location, velocity, multipath scenario, fading, and interference from other base stations.

7.3 References

[7.1] F. T. Ulaby, R. K. Moore, and A. K. Fung. *Microwave Remote Sensing, Volume 1*, Addison-Wesley, 1981.

[7.2] TIA/TR45.5.3.1/95.01.10.05 and TR45.5.1.3/95.01.11.04. Reverse Link Simulation Results, January 1995.

[7.3] A. J. Viterbi, A. M. Viterbi, K. S. Gilhousen, and E. Zehavi. Soft Handoff Extends CDMA Cell Coverage and Increases Reverse-link Capacity, *IEEE J. Select. Areas Commun.*, vol. 12, pp. 1281-1288, October 1994.

[7.4] ANSI J-STD-008. *Mobile Station - Base Station Compatibility Requirements for 1.8 and 2.0 GHz CDMA PCS,* March 1995.

[7.5] TIA/TR45.5.1.3/95.03.14.03. Forward-link Simulation Results, March 1995.

[7.6] TIA/TR45.5.1/95.07.18._. Range Vs. Number of Subscribers for the Forward and Reverse-links (Rev. 2), July 1995.

CHAPTER 8

Capacity

In CDMA, all subscribers use the same carrier. Moreover, each signal is coded to appear as noise (interference) with respect to every other signal. Accordingly, each signal is contained in a background of broadband interference generated by other users. A balance between maintaining call integrity and restricting interference levels is maintained by controlling the power of each signal so that it arrives at its intended receiver with the minimum required signal-to-interference ratio (SIR) level.

Either transmit power constraints or the system's self-generated interference ultimately restricts CDMA system capacity. On the reverse link, the system reaches capacity when a mobile station has insufficient power to overcome interference from all the other mobile stations. On the forward link, air-interface capacity is reached when no additional fractional power is available to add an additional user, that is forward-link power limited capacity is reached when the total transmitted power required to successfully maintain all users hosted by a base station exceeds the base station's restrictions. The power needed on either link is fundamentally related to E_b/I_o requirements at the intended receiver. Capacity can vary because the requirement varies with changing conditions. Mobile station soft handoff populations can also impact capacity.

The following section considers limiting factors on CDMA reverse- and forward-link capacities in more detail. We address reverse- and forward-link capacities separately. Chapter 10 includes numerical information on capacity (that is, Erlangs with blocking).

8.1 Reverse-Link Capacity

To place a call, a CDMA mobile station must have sufficient power to overcome the interference generated by all other CDMA mobile stations within the band, that is, the received signal at the base station must achieve a required SIR. The mobile station transmit power required at any given time will depend on the path loss from the mobile station to the base station and the total level of reverse-link interference. The latter primarily depends on the number and position of other CDMA mobile stations.

The establishment of an additional call raises the interference levels seen by all mobile stations, and each mobile station must appropriately increment its transmit power accordingly to maintain call integrity. This process repeats itself with each additional mobile station until it reaches a limit. At this limit, a new mobile station, regardless of position, does not have enough power to overcome the level of interference generated by current mobile stations and the current mobile stations do not have enough power to overcome the additional interference a new call would generate. This limit represents an upper bound on system capacity.

The limit occurs because mobile station units eventually have insufficient power to achieve the required SIR at the base station. Accordingly, any factor that varies the level of signal and/or interference at the base station has an influence on this limit. For example, a heavily loaded neighbor cell will increase the level of interference and lower the base station capacity. The amount of reverse-link voice activity will impact capacity because the mobile station restricts the output power when the user is not speaking.

The many factors influencing CDMA capacity give rise to a desirable flexibility in system operation. The dependence on interference levels means that a cell's capacity is inherently dynamic, that is, a base station can naturally absorb more users if neighboring cells are lightly loaded. In addition, the system can naturally exploit the reduced levels of interference generated by low voice activity.

This flexibility also makes it difficult to assess CDMA capacity in a manner that will be applicable to all situations. A means to obtain a useful reference point is by assessing the number of allowed calls in a centrally embedded base station under ideal conditions. This point is computed by assuming that power control acts to maintain a constant, minimal E_b/I_o at the base station receiver for each reverse-link signal, that is,

$$\left(\frac{E_b}{I_o}\right)_{\text{at the receiver}} = \left(\frac{E_b}{I_o}\right)_{\text{required}} \equiv d \,. \tag{8.1}$$

In (8.1), E_b/I_o is the ratio of bit energy to total noise power spectral density. This quantity represents a standard figure of merit for digital modulation schemes. The bit energy

Reverse-Link Capacity

is obtained by dividing the received signal power by the bit rate (see, also, (7.3)). The total noise power spectral density is obtained by summing background thermal noise density, N_{th}, and the spectral density of broadband interference from all other CDMA users. The former must be adjusted by the base station noise figure F. The latter is composed of contributions from users both within the cell and in other cells.

It must be realized that different categories of users have different E_b/I_o requirements to maintain a certain FER. For example, static users need less E_b/I_o to maintain an FER of one percent than users traveling at low speeds (for example, at 3 mph) and higher mobility users (about 30 mph) have a different requirement E_b/I_o compared to low mobility and static cases.

The product of thermal noise density N_{th}, CDMA bandwidth W, and the base station noise figure F, will be called the *base station noise*. This is the minimum system noise (thermal plus receiver generated) at the receiver input, and it can also be considered as the noise floor of the system. This is the amount of noise that a single user would have to overcome. It is a useful reference point for measuring the strength of incoming signals.

The range of required E_b/I_o values at the cell-site receiver is a slowly varying function of mobile station speed and multipath condition. The number of paths that can be separately demodulated at the (Rake) receiver determines the latter. Generally, a minimum of two multipaths can be assumed since the diversity receive antennas employed at the cell site guarantee the presence of at least two paths. The narrow range of values within the two multipath case permit the use of a worst-case value (7 dB) for all mobile stations without being overly conservative.

Within the cell, the restriction of equal E_b/I_o for all calls can be shown to require that all signal strengths received at the cell site are equal to the common term S. For any mobile station, the in-cell interference is then $(N - 1)S$, where N is the number of mobile stations within the cell. This term is the primary source of interference on the reverse-link.

The co-channel interference from mobile stations outside the cell is a secondary source of interference and can be taken to be a fraction β of the in-cell interference. The low transmitter strength and increased distance (path loss) of the surrounding mobile stations produce an interference level that can be typically characterized by $\beta < 1$ and that tends to be below the cell-site noise level. Unlike the in-cell interference, the interference from mobile stations outside the cell is not under power control by the cell-site receiver and is therefore more difficult to determine; however, only the aggregate effect of all outside mobile stations need be known with any accuracy. The large number of surrounding mobile stations, as well as the inherent randomness in their locations, generates an averaging effect that facilitates prediction.

The relatively low level of co-channel interference allows the use of a β corresponding to surrounding fully loaded cells without being overly conservative. Furthermore, for large N, all interference can be reduced by a factor α that reflects the mean voice activity (less than one hundred percent) across all active channels on the reverse-link. The derivation presented below follows the development in Reference [8.1]. All of the above considerations are detailed further in Reference [8.1] and lead to

$$\frac{E_b}{I_o} = \frac{S/R}{FN_{th} + \frac{\alpha(1+\beta)(N-1)S}{W}} = G\frac{S}{FN_{th}W + \alpha(1+\beta)(N-1)S}, \quad (8.2)$$

where

E_b = bit energy,
I_o = power spectral density of thermal noise plus interference,
F = base station noise figure,
N_{th} = power special density of thermal noise,
S = received signal strength,
R = bit rate,
α = voice activity factor,
β = interference factor,
N = number of mobile stations in cell,
W = system bandwidth, and
$G = W/R$ = processing gain.

As shown, the required E_b/I_o can be expressed as the product of processing gain and SIR, where the interference is the sum of the base station noise and interference from other CDMA users. The expression (8.2) can be rewritten to explicitly indicate the number of mobile station calls N:

$$N = \frac{G}{\alpha d(1+\beta)} + 1 - \frac{1}{\alpha(1+\beta)}\frac{FN_{th}W}{S}$$

$$\Rightarrow N_{max} = \frac{G}{\alpha d}\frac{1}{(1+\beta)} + 1, \quad (8.3)$$

where d is the required E_b/I_o as defined in (8.1) and (8.2).

Table 8–1 Values Used for Evaluating the Pole-Point

Parameter	Definition	Value
$G = W/R$	processing gain	85.333 (19.3 dB)
α	mean voice activity factor	0.4
β (for an omni-cell)	omnidirectional cell interference factor	0.6
d	required E_b/I_o	5 (7 dB)
β (for a single sector of a sectored cell)	other cells/sectors interference factor	0.85

Table 8–2 Receiver Interference Margin and Number of Channels

Receiver Interference Margin		Number of Channels	
in dB	Loading (percent of pole point)	OMNI 27 (pole point)	Three-Sectored 24 (pole point)
0 dB	0.0%	0	0
1 dB	20.6%	5	5
2 dB	37.0%	10	9
3 dB	50.0%	13	12
3.4 dB	55.0%	14	13
3.8 dB	58.3%	15.7	14
4.6 dB	65.0%	17	15

In (8.3), the finite limit on capacity can be conveniently reached by letting the signal-to-cell-site noise ratio go to ∞ (that is, by letting the received signal power become unbounded with respect to the cell-site noise). This capacity is called the *pole point* or *power pole* and represents a theoretical maximum that cannot be reached but which serves as a useful reference point. Cell loading can be conveniently expressed as a fraction of the pole point. With appropriate choice of β, the pole (8.3) applies to an omnidirectional cell or a single sector of a directional cell.

Example: The pole point (8.3) may be evaluated for typical values of the parameters used. The values used to evaluate the pole point are tabulated in Table 8–1.

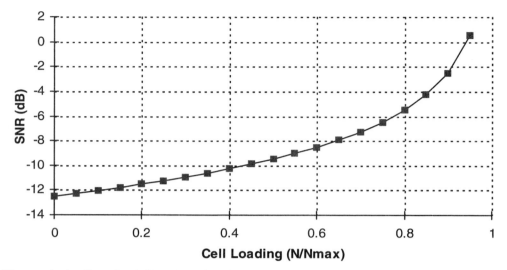

Figure 8–1 Required Signal-to-Cell-Site-Noise-Power Ratio Per User as Function of Cell Loading (No Soft Handoff)

Note: Base station noise is thermal noise adjusted by the base station noise figure.

The pole point obtained using these values is approximately 24 for a sectored cell, and 27 for an omni-cell using the 13 Kbps vocoder option as listed in Table 8–2. ❑

The pole point is inversely proportional to the required E_b/I_o ratio. The reverse-link capacity limit in CDMA is thus *soft* rather than *hard*, and system capacity can be increased by lowering the value of E_b/I_o and slightly degrading the quality of all calls. This feature could be useful in temporarily handling unusually heavy traffic and/or in avoiding dropped calls.

We can obtain further insight into capacity dynamics by rewriting the expression (8.3) to obtain the average required signal power per user as

$$\frac{S}{FN_{th}W} = \frac{1}{N_{max}\alpha(1+\beta)(1-\frac{N}{N_{max}})} \cdot \quad (8.4)$$

Figure 8–1 plots this relationship. The required signal-to-cell-site-noise-power ratio per user rises in a non-linear fashion with cell loading. Accordingly, the average power *cost* per additional mobile is much greater for a heavily loaded cell; moreover, a cell that is too heavily loaded will be unable to accommodate statistical fluctuations in the aggregate voice activity of its members. Operating below the point at which the non-linear slope becomes too steep (for example, fifty to sixty percent) can roughly set appropriate capacity limits.

The expression (8.4) can be further adjusted to reflect the total received power (interference) $P_{rec} = \alpha(1 + \beta)NS$. The ratio of P_{rec} to the base station noise can be expressed entirely in terms of the cell loading $\mu = N/N_{max}$, and indicates that the interference also rises in a non-linear fashion with cell loading:

$$\frac{P_{rec}}{FN_{th}W} = \frac{\mu}{1-\mu}. \qquad (8.5)$$

This relationship between power and loading can also be expressed using the ratio of total power (received power plus base station noise) to base station noise:

$$\frac{P_{total}}{FN_{th}W} = \frac{P_{rec} + FN_{th}W}{FN_{th}W} = \frac{1}{1-\mu}. \qquad (8.6)$$

Noting that the ratio of total power to base station noise doubles every time half of the remaining pole capacity is used succinctly summarizes the non-linear behavior of this curve. For example, the ratio rises from 0 to 3 dB when loading increases from 0 to 0.5 and rises from 3 dB to 6 dB when loading rises from 0.5 to 0.75.

Observe that the reverse-link budget in Table 7–1 uses a loading factor of fifty-five percent. For the parameters assumed in Table 8–1, a pole loading of fifty-five percent corresponds to thirteen users per sector and a rise over base station noise of 3.4 dB (from (8.6)). Figures 8–2a and 8–2b, respectively, plot the relationship of (8.5) and (8.6).

Expression (8.2) also provides a convenient means of assessing the impact of external (that is, non-CDMA) interference sources. The received power I_{ext} of external interference sources can be incorporated into (8.2) by noting that the base station receiver's decoding (de-spreading) operation would spread narrowband external interference across the CDMA bandwidth. We can therefore assess the cost of external interference by translating its impact into an equivalent rise in the base station noise figure.

8.2 Forward-Link Capacity

Upper limits on forward-link capacity are fundamentally determined by restrictions on sector-radiated power and by mobile receiver E_b/I_o requirements. The forward-link signal comprises message traffic for mobiles, a sector-specific signal (pilot) used for sector

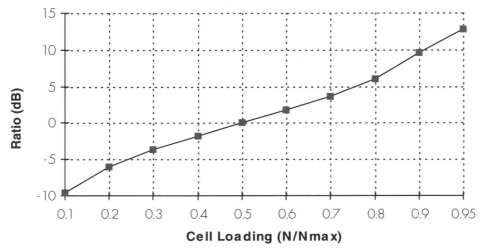

(a) Ratio of Base Station Received Power (Interference) from all Mobile Stations to Base Station Noise as Function of Cell Loading

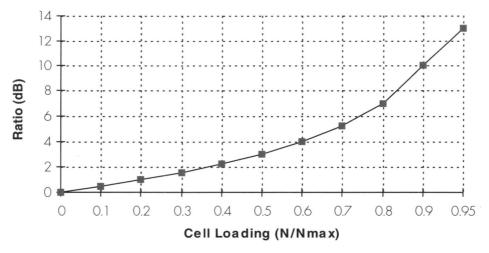

(b) Ratio of Total Power (Interference Plus Base Station Noise) to Base Station Noise as a Function of Cell Loading

Figure 8–2 Plot of Equations (8.5) and (8.6)

acquisition and tracking, and miscellaneous signals (for example, sync, paging). The base station allocates total power among these functions. Capacity limits are reached when the sum of the allocations required to meet E_b/I_o requirements reaches one hundred percent of the available transmit power.

To maintain voice quality, the sector transmitter must achieve the required E_b/I_o at each mobile within the serving sector. The channel impairment at the mobile receiver consists of thermal noise, co-channel interference from links servicing other mobiles within the serving sector, and co-channel interference from links servicing mobiles outside the serving sector. The last is essentially the power broadcast by surrounding CDMA sectors and is typically the primary source of co-channel interference.

The co-channel interference composed of signals intended for other mobiles within the sector is coded to be orthogonal to the mobile signal and can be screened out effectively. (Multipath may impair the orthogonality). The signals from links servicing mobiles outside the sector are simply seen as broadband interference. Power control algorithms enable the mobile to request the level of power it requires to overcome the total interference, which varies with mobile position. The sector responds to the mobile's request by adjusting the power allocated to the mobile's traffic.

The requirement that a generous fraction of power must consistently be allocated to the sector pilot restricts these adjustments. All mobiles use this signal in acquiring the cell, demodulating forward-link signals, and making handoff decisions. Thus, only some eighty percent of sector power will typically be available for message traffic. Capacity limits are reached when this power, distributed among sector users, is insufficient to achieve E_b/I_o requirements at the mobile receivers. This limit will depend on mobile position with respect to its serving cell as well as with respect to its neighbor cells.

Unlike the reverse link, the range of required E_b/I_o values in the forward link is a strongly varying function of mobile speed and multipath condition; moreover, the mobile receiver does not employ antenna diversity. The latter difference means that a minimum number of two paths cannot be guaranteed unless the mobile is known to be in a soft and/or softer handoff state. Accordingly, the required value of E_b/I_o can vary considerably from mobile to mobile. This variation, coupled with the inherent randomness in mobile location as well as in the interference level from surrounding cells, heightens the complexity of forward-link capacity analysis.

In spite of these difficulties, considerable insight into the behavior of the forward link can be gained by examining the fundamental relation expressing the fractional allocation of total sector power assigned to a traffic channel:

$$x = \frac{d}{G} \frac{FN_{th}W + P_{r,same} + P_{r,other}}{P_{r,host}} \approx \frac{d}{G} \frac{P_{r,other}}{P_{r,host}}, \tag{8.7}$$

where

x = fractional traffic power allocation for mobile receiver,
d = E_b/I_o requirement,
G = spread spectrum processing gain (bandwidth normalized by data rate),
$P_{r,host}$ = received total power at mobile receiver from host sector,
$P_{r,same}$ = received total power at mobile receiver from other same-sector traffic links, and
$P_{r,other}$ = received total power at mobile receiver from surrounding sector.

Relation (8.7) indicates that the fractional allocation of total sector power to a traffic channel is determined by two ratios: that of E_b/I_o requirement to processing gain and that of other-sector interference to same(host)-sector interference. To support a large number of mobiles, the allocation x must be small; accordingly, the E_b/I_o requirement d should be small with respect to the processing gain and/or the other-sector interference should be small with respect to the host power.

The relation (8.7) imposes restrictions on system performance. The position of the mobile receiver within the sector footprint strongly influences the ratio of other- to same-sector interference. Ideally, the receiver E_b/I_o requirements should be small enough with respect to G to alleviate this dependence; however, the wide range of possibilities for d may include values that are comparable to the processing gain.

These circumstances can limit the achievement of high E_b/I_o values with small x to receivers experiencing favorable interference environments. Such situations may arise due to geometrical and/or loading asymmetries but cannot generally be created by reducing the nominal cell radius. Uniform reduction in an interference-limited environment will not alter the relation (8.7) because the ratio of other- to same-sector interference will remain unchanged. Similarly, the availability of additional power at the host will not necessarily improve performance if comparable amounts of additional power are broadcast by the surrounding sectors.

The strong dependence of forward-link performance on many underlying factors prevents the simplification that was possible in the reverse-link analysis: the use of a worst case in design would produce an overly conservative result. The preferred approach is to assess many random combinations of the contributing factors in order to ascertain whether the forward link can support mobile operations within the nominal footprint and at the number of channels generated by the reverse link.

Analyses/simulations that comprise multiple cell sites in which the location, fading, speed, voice activity, and multipath state of mobiles are randomly chosen are generally the means for considering such assessments. (In some cases, the selection process is weighted by system performance constraints, for example, a mobile in a softer handoff state is guaranteed two paths.) In these computations, the number of mobiles placed in each sector is variable and determined by an Erlang B model driven by the sector's traffic load. This method accurately simulates operation and allows the CDMA forward link to properly exploit the reduced interference background presented by surrounding cells that are not likely to all be simultaneously at a blocking state (that is, the state where all available channels are occupied). The forward link achieves its performance by sharing the allowed interference level across sectors, that is, the channels available in the air interface can be viewed as being pooled across sector boundaries. Such analyses indicate that the forward link is generally viable within the reverse-link footprint provided that the E_b/I_o requirements at the mobile receiver are restricted to a suitable set of possibilities.

8.3 Reference

[8.1]   K. S. Gilhousen, I. M. Jacobs, R. Padovani, A. J. Viterbi, L. A. Weaver, Jr., and C. E. Wheatley. On the Capacity of a Cellular CDMA System, *IEEE Trans. Vehicular Technology*, vol. 40, pp. 303-312, May 1991.

CHAPTER 9

Coverage

The nature of CDMA is that it interrelates coverage, capacity, and interference considerations. A CDMA traffic signal must have sufficient level to reach the required signal-to-interference ratio at its intended receiver. Accordingly, the area in which CDMA coverage can be achieved depends on the level of interference that must be overcome. Since all CDMA signals share the same carrier, the interference is primarily self-generated and depends on the traffic in host and neighbor cells. External (that is, non-CDMA) sources of interference will contribute to the level as well. The extent of the coverage area also depends on the distance and thus path loss between the transmitter and the receiver in addition to the distance (path loss) between the receiver and the interference sources. These considerations apply to both the forward- and reverse-links.

9.1 Reverse-Link Coverage Area

On the reverse link, the area of coverage is comprised of those locations from which a mobile station has sufficient transmit power to overcome the total level of interference at the base station. A mobile station's limited transmit power will therefore restrict the mobile-station-to-base-station path loss that can be tolerated. This path loss can be equated to a maximum mobile-station-to-base-station distance. (In a typical anisotropic loss situation, this equivalence will vary with direction.) Since a heavily loaded cell exhibits high levels of interference, this maximum distance will shrink as the cell loading increases. High user populations in neighbor cells and external sources of interference will reduce the distance as well. The permitted distance will also vary with the mobile station's available transmit power (mobile station class).

Computing the maximum allowed distance (for the reverse-link) under idealized conditions of power control and path loss allows for some insight into these considerations. These computations make use of terms and observations introduced in Section 8.1. The required mean signal strength is a function of cell loading and of external interference, and it can be written as (derived from Equation (8.2))

$$\overline{S}(N) = (FN_{th}W)\frac{1}{\alpha(1+\beta)(N_{max}-N)}\left(1 + \frac{I_{ext}}{FN_{th}W}\right), \qquad (9.1)$$

where

$\overline{S}(N)$ = required received signal strength for N mobile stations,
F = base station noise figure,
N_{th} = power spectral density of thermal noise,
W = system bandwidth,
α = voice activity factor,
β = interference factor,
N_{max} = pole point,
N = number of mobile stations, and
I_{ext} = received external interference power.

$\overline{S}(N)$ gives us the required average power, and it is written as a function of N to stress its dependence on cell loading. However, as we are interested in coverage predictions in this section, variations in the received power have to be taken into account. As detailed in the reverse-link budget calculations of Chapter 7, after providing for fading under soft handoff, an additional link margin of M dB is required to ensure that the signal strength is adequate to provide for coverage ninety percent of the time at the cell edge. At any other point inside the cell, the mobile station's received signal at the base station will meet the SIR requirement for more than ninety percent of the time. Thus the received signal strength, $\overline{S}(N)$, given in (9.1), has to be augmented by M dB. The mobile station's maximum transmit power P_m^t, adjusted for the mean propagation loss, mobile station antenna gain, base station antenna gain, and the M dB extra link margin, must achieve the value $\overline{S}(N)$. That is, the power received at the base station must be larger than $M \times \overline{S}(N)$. If path loss is modeled by an intercept loss l_0 and a 10γ dB/decade rolloff (that is, the path loss is proportional to $l_0 \times r^\gamma$), this requirement becomes

Reverse-Link Coverage Area

$$r \leq \left(\frac{P_m^t \times G_{\text{mobile station}} \times G_{\text{base station}}}{l_0 \times M \times \overline{S}(N)} \right)^{1/\gamma}, \qquad (9.2)$$

where

- r = distance from mobile station to base station (maximum value of r is the radius of reverse-link coverage area),
- l_0 = one-mile intercept loss,
- $G_{\text{mobile station}}$ = mobile station gain versus omnidirectional source,
- $G_{\text{base station}}$ = base station gain versus omnidirectional source,
- P_m^t = maximum mobile transmit power,
- γ = path loss exponent,
- $\overline{S}(N)$ = required received signal strength for N mobile stations, and
- M = additional link margin to account for fading under soft handoff.

Note that (9.2) is a restatement of (7.10).

Observe that the maximum distance increases with mobile station transmitter strength, but decreases with cell loading and strength of external interference. These relations are non-linear; for example, if path loss is 35.2 dB/decade, the distance varies as the 1/3.52 power of these quantities.

Example: We plot the maximum value of r as determined by Equation (9.2) in the absence of external interference as a function of cell loading in Figure 9–1. We obtained this curve using the values listed in Table 9–1. The one-mile intercept point and the path loss exponent are based on the COST 231 Hata model. It should be noted that Figure 9–1 is just an illustrative example based on the propagation path loss model stated above and for the parameters given in Table 9–1. It should not be used as a reference number for actual deployment studies. Service providers should use the appropriate propagation model to size base stations. ❑

Note that loss in coverage caused by external interference can be examined by noting that the impact of external interference can be equated to a rise in the base station noise figure (see Section 8.1).

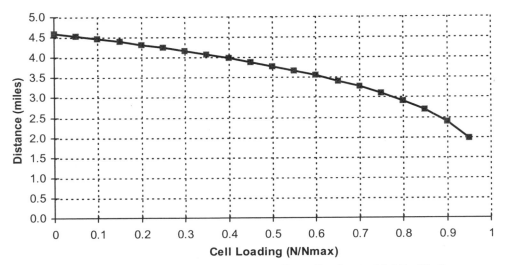

Figure 9–1 Maximum Allowed Distance from Base Station to Mobile Station

Table 9–1 Values Used in Computing Radius of Coverage

Parameter	Definition	Value
N_{th}	power spectral density of thermal noise	–174 dBm/Hz
W	system bandwidth	1.23 MHz
F	base station noise figure	5.0 dB
α	voice activity factor	0.4
β	interference factor	0.85
$G_{\text{mobile station}}$	mobile station gain minus body loss	0 dB
$G_{\text{base station}}$	base station gain	13 dBi
l_0	path loss at one-mile intercept	130.6 dB
P_m^t	maximum mobile station transmit power	200 mW
γ	path loss exponent	3.52
M	fading margin under soft handoff	6.3 dB

9.2 Forward-Link Coverage Area

On the forward link, the area of coverage comprises those locations where sufficient power to overcome interference can be allocated to a mobile station. This interference is composed of in-cell traffic (power allocated to other mobile stations within the cell) and power received from other base stations. Power allocated to in-cell traffic can be screened out with some success via coding known to base station members; however, multipath effects impair this process.

In general, base station output power limitations will ultimately restrict the areas in which sufficient power can be supplied to a mobile station. These areas depend on the propagation loss from the mobile station to the base station as well as the propagation loss from the mobile station to the interferers. Coverage is also dependent on cell loading and traffic distribution in the sense that the available power per mobile station drops with the increase in base station loading and that power requirements for mobile stations distributed close to the serving base station are likely to be more modest than those for mobile stations distributed near the cell edges.

In the remainder of this chapter, we will consider the forward-link coverage in terms of the probability that the mobile station can detect the CDMA pilot. The following examines two cases with hard and soft handoff.

9.2.1 Coverage Probability for Pilot Channel

Prior to CDMA mobile station access, the strongest pilot selected among several CDMA base stations has to be acquired first. The pilot channel also aids the handoff operation. When in soft handoff, a CDMA mobile station continuously scans for the pilot channels transmitted by each base station and maintains communication simultaneously with multiple base stations whose power exceeds a given threshold. This section examines the coverage probability of the CDMA pilot with hard and soft handoff. Specifically, for hard and soft handoff, based on the detailed derivation in Section 9.3, we examine the probability that the ratio of the received pilot chip energy to spectral densities of the thermal noise plus the interference is above the required threshold. We also present numerical results and trade-offs among the system parameters such as the CDMA pilot power percentage allocation, the required E_c/I_o threshold and the use of hard and soft handoff as well as the coverage probability.

Let δ be the pilot channel E_c/I_o threshold needed to achieve certain coverage probability. As shown in Section 9.3, the coverage probability at location (r, θ) is

$$\Pr\left\{\frac{E_{c,\text{pilot}}}{I_o} \geq \delta\right\}$$

$$= \int_{10\log_{10}\frac{\eta}{\left(1+\frac{1}{\delta}\right)\mu}}^{\infty} \left\{1 - Q\left(\frac{10\log_{10}\eta + 10\log_{10}\left[\left(1+\frac{1}{\delta}\right)\frac{\mu}{\eta}10^{0.1x} - 1\right] - \varepsilon_Y}{\sigma_Y}\right)\right\} p_X(x)\,dx. \quad (9.3)$$

In (9.3), μ is the fraction of base station power for pilot. The variables ε_Y and σ_Y^2 are mean and variance, respectively, of the sum of in- and outer-cell shadowing. In addition, η is the inverse of the CDMA carrier to thermal noise power ratio at the cell boundary that is defined as

$$\eta = \frac{FN_{\text{th}} W L_b^{Tx} L_m^{Rx} 10^{0.1[10\gamma \log_{10} R_c + 10\log_{10} l_0]}}{P_b G_b^{Ant,Tx} G_m^{Ant,Rx}},$$

in which

P_b = power per CDMA carrier at base station transmit filter input,
L_b^{Tx} = losses due to transmit filter, transmission lines, and cables at the transmit side of the base station,
$G_b^{Ant,Tx}$ = transmitter antenna gain of the base station,
$G_m^{Ant,Rx}$ = receiver antenna gain of the mobile station,
L_m^{Rx} = losses due to receive filter, transmission lines, and cables at the receiver side of the mobile station,
γ = propagation path loss exponent,
l_0 = path loss at intercept,
F = receiver noise figure of the mobile station,
N_{th} = thermal noise power spectral density,
W = CDMA signal bandwidth, and
R_c = cell radius.

The probability density function $p_X(x)$ is

$$p_X(x) = \sum_{j=1}^{J} \frac{1}{\sqrt{2\pi\sigma^2}} \exp\left[-\frac{(x-\varepsilon_j)^2}{2\sigma^2}\right] \prod_{\substack{i=1 \\ i \neq j}}^{J} \left[1 - Q\left(\frac{x-\varepsilon_i}{\sigma}\right)\right],$$

with ε_j and σ^2 being the mean propagation loss (dB) between the mobile station to the j-th base station and its variance, respectively.

For numerical evaluation of the resulting analytical expressions, we assume there are thirty-seven hexagonal cells; the mobile station under consideration is in the centered cell surrounded by thirty-six cells, as shown in Figure 9–2. The lognormal shadowing channels are assumed to have propagation exponent $\gamma = 4$ and standard deviation $\sigma = 8$ dB. The lognormal shadowing between the mobile station and different base stations are un-correlated. We examine hard handoff and two-way and three-way soft handoff. We assume there are 240 possible mobile station locations in a 120-degree sector, each location separated by five degrees and one-tenth of the cell radius. We calculate coverage probability for each location. Because of symmetry, it is sufficient in the case of an omni-cell to calculate those locations in only a 120-degree sector. For a sectored cell, we assume an ideal antenna pattern so that there is no interference from adjacent sectors. Every cell or sector uses the same RF bandwidth.

We calculate the distribution of the location coverage probabilities weighted by their associated bin areas. We compute area coverage probability as the average of the location coverage probabilities weighted by the associated bin areas. We also compute cell boundary coverage probability as the average of the location coverage probabilities on the circular boundary; thus it gives the coverage probability for mobile stations moving on the cell boundary. We assume pilot power percentage allocation to be from –10 dB to 0 dB, and E_c/I_o threshold to be from –15 dB to –20 dB. These numerical results are plotted in Figures 9–3 to 9–9 for hard handoff, in Figures 9–10 to 9–16 for two-way soft handoff, and in Figures 9–17 to 9–23 for three-way soft handoff. Figures 9–24 to 9–29 summarize comparisons of the required pilot power percentage allocation between hard handoff and soft handoff.

Numerical results indicate that the trade-off can be performed among the system parameters such as pilot power percentage allocation, E_c/I_o threshold, and coverage probability. Note that in the shadowing environment, the diversity gain provided by soft handoff reduces the required pilot power allocation as compared to that required for hard handoff. For the same area coverage probability, the use of soft handoff reduces the required pilot power by 4 dB to 5 dB compared with that for hard handoff. For the same cell boundary coverage probability, the use of soft handoff reduces the required pilot power by 7 dB to 8 dB as compared with that for hard handoff. With soft handoff, suitable choices of pilot power percentage allocation and E_c/I_o threshold can achieve the target area coverage probability (for example, ninety percent) or the target cell boundary coverage probability (for example, ninety percent).

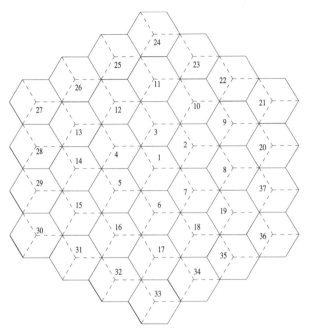

Figure 9–2 Hexagonal Layout for Thirty-Seven Cells

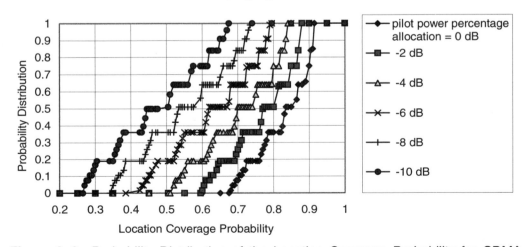

Figure 9–3 Probability Distribution of the Location Coverage Probability for CDMA Pilot with Hard Handoff for E_c/I_o Threshold −15 dB and Pilot Power Percentage Allocation from −10 dB to 0 dB

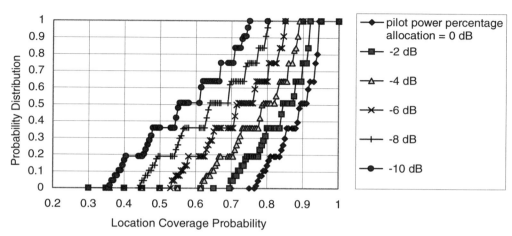

Figure 9–4 Probability Distribution of the Location Coverage Probability for CDMA Pilot with Hard Handoff for E_c/I_o Threshold −17.5 dB and Pilot Power Percentage Allocation from −10 dB to 0 dB

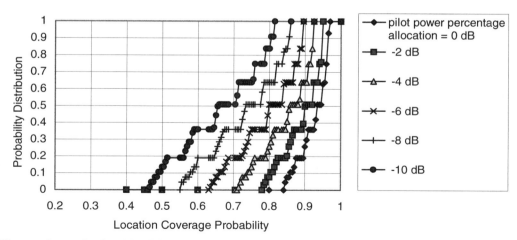

Figure 9–5 Probability Distribution of the Location Coverage Probability for CDMA Pilot with Hard Handoff for E_c/I_o Threshold −20 dB and Pilot Power Percentage Allocation from −10 dB to 0 dB

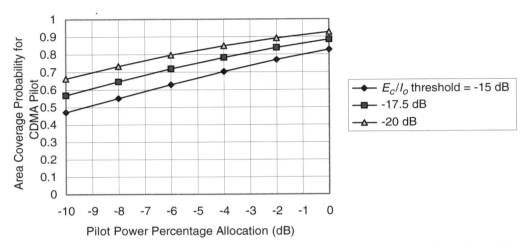

Figure 9–6 Area Coverage Probability for CDMA Pilot with Hard Handoff as a Function of Pilot Power Percentage Allocation with E_c/I_o Threshold −15, −17.5, and −20 dB

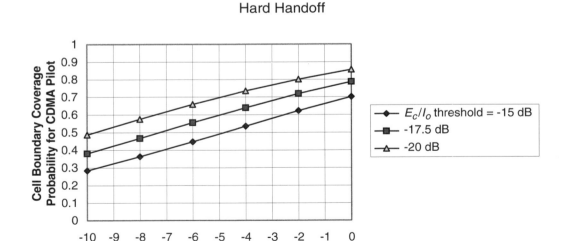

Figure 9–7 Cell Boundary Coverage Probability for CDMA Pilot with Hard Handoff as a Function of Pilot Power Percentage Allocation with E_c/I_o Threshold −15, −17.5, and −20 dB

Forward-Link Coverage Area

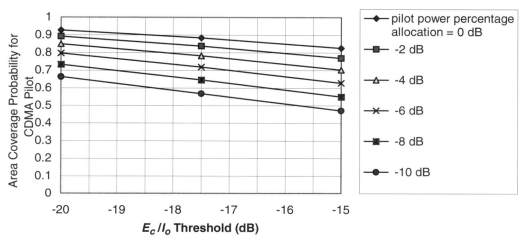

Figure 9–8 Area Coverage Probability for CDMA Pilot with Hard Handoff as a Function of E_c/I_o Threshold for Pilot Power Percentage Allocation from −10 to 0 dB

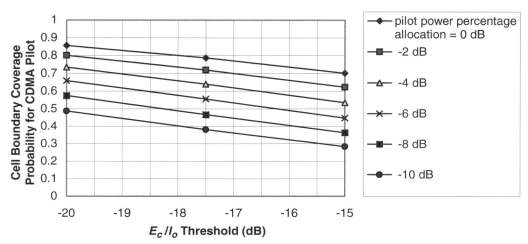

Figure 9–9 Cell Boundary Coverage Probability for CDMA Pilot with Hard Handoff as a Function of E_c/I_o Threshold for Pilot Power Percentage Allocation from −10 to 0 dB

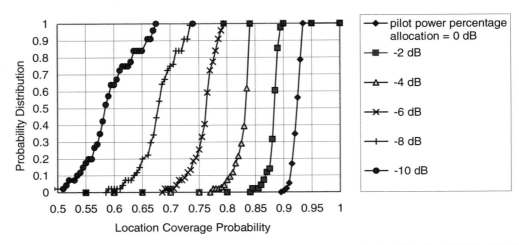

Figure 9–10 Probability Distribution of Location Coverage Probability for CDMA Pilot with Two-Way Soft Handoff with E_c/I_o Threshold –15 dB and Pilot Power Percentage Allocation from –10 to 0 dB

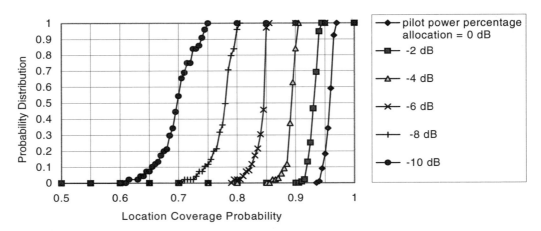

Figure 9–11 Probability Distribution of Location Coverage Probability for CDMA Pilot with Two-Way Soft Handoff with E_c/I_o Threshold –17.5 dB and Pilot Power Percentage Allocation from –10 to 0 dB

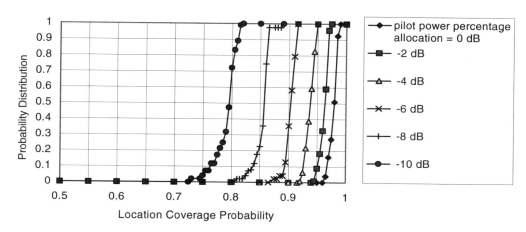

Figure 9–12 Probability Distribution of Location Coverage Probability for CDMA Pilot with Two-Way Soft Handoff with E_c/I_o Threshold –20 dB and Pilot Power Percentage Allocation from –10 to 0 dB

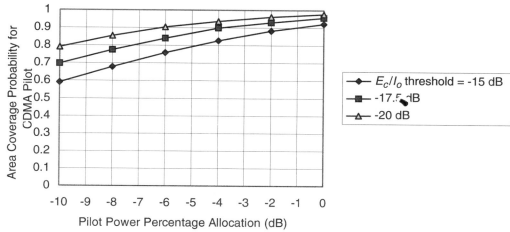

Figure 9–13 Area Coverage Probability for CDMA Pilot with Two-Way Soft Handoff as a Function of Pilot Power Percentage Allocation with E_c/I_o Threshold –15 dB, –17.5 dB, and –20 dB

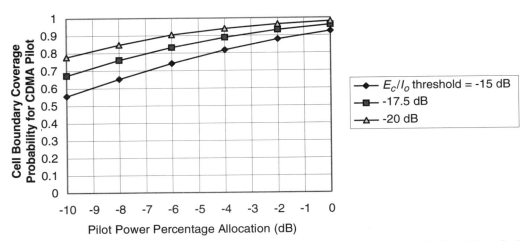

Figure 9–14 Cell Boundary Coverage Probability for CDMA Pilot with Two-Way Soft Handoff as a Function of Pilot Power Percentage Allocation with E_c/I_o Threshold −15, −17.5, and −20 dB

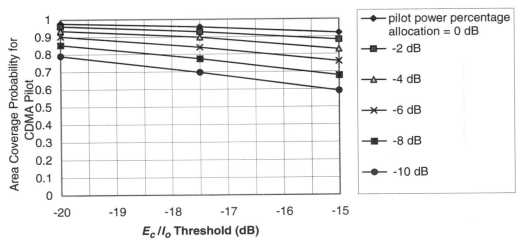

Figure 9–15 Area Coverage Probability for CDMA Pilot with Two-Way Soft Handoff as a Function of E_c/I_o Threshold for Pilot Power Percentage Allocation from −10 to 0 dB

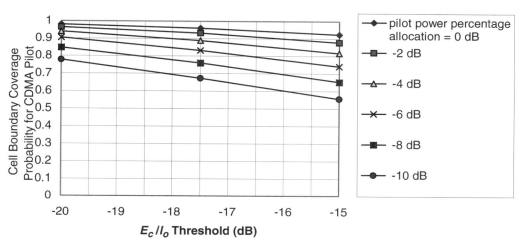

Figure 9–16 Cell Boundary Coverage Probability for CDMA Pilot with Two-Way Soft Handoff as a Function of E_c/I_o Threshold for Pilot Power Percentage Allocation from −10 to 0 dB

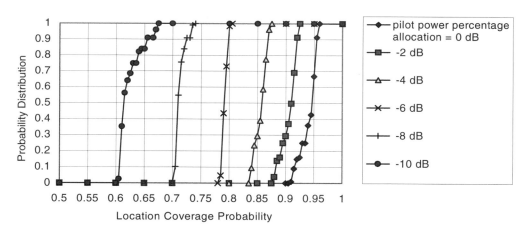

Figure 9–17 Probability Distribution of Location Coverage Probability for CDMA Pilot with Three-Way Soft Handoff with E_c/I_o Threshold −15 dB and Pilot Power Percentage Allocation from −10 to 0 dB

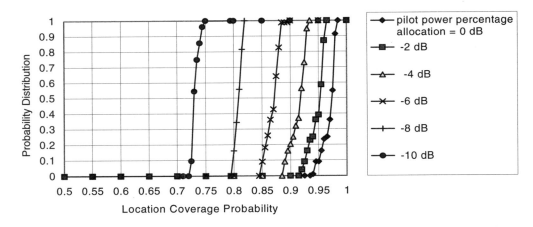

Figure 9–18 Probability Distribution of Location Coverage Probability for CDMA Pilot with Three-Way Soft Handoff with E_c/I_o Threshold −17.5 dB and Pilot Power Percentage Allocation from −10 to 0 dB

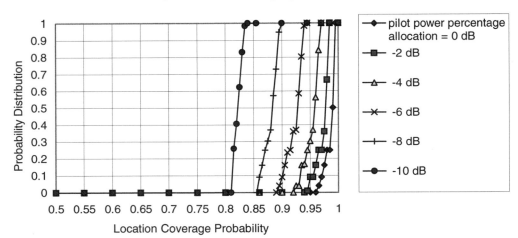

Figure 9–19 Probability Distribution of Location Coverage Probability for CDMA Pilot with Three-Way Soft Handoff with E_c/I_o Threshold −20 dB and Pilot Power Percentage Allocation from −10 to 0 dB

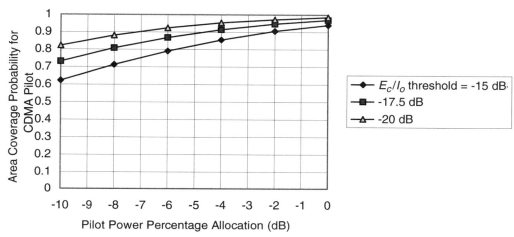

Figure 9–20 Area Coverage Probability for CDMA Pilot with Three-Way Soft Handoff as a Function of Pilot Power Percentage Allocation with E_c/I_o Threshold −15 dB, −17.5 dB, and −20 dB

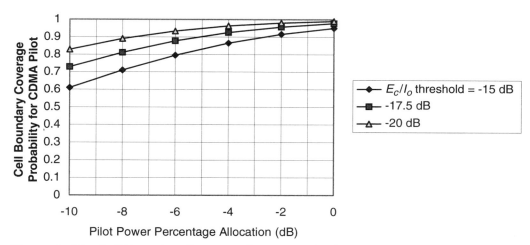

Figure 9–21 Cell Boundary Coverage Probability for CDMA Pilot with Three-Way Soft Handoff as a Function of Pilot Power Percentage Allocation with E_c/I_o Threshold −15 dB, −17.5 dB, and −20 dB

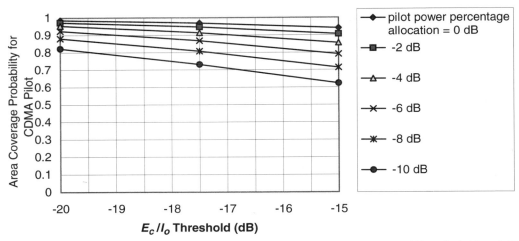

Figure 9-22 Area Coverage Probability for CDMA Pilot with Three-Way Soft Handoff as a Function of E_c/I_o Threshold for Pilot Power Percentage Allocation from −10 to 0 dB

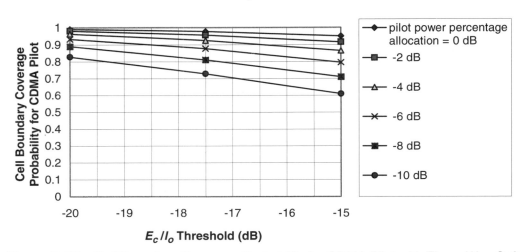

Figure 9-23 Cell Boundary Coverage Probability for CDMA Pilot with Three-Way Soft Handoff as a Function of E_c/I_o Threshold for Pilot Power Percentage Allocation from −10 to 0 dB

Forward-Link Coverage Area

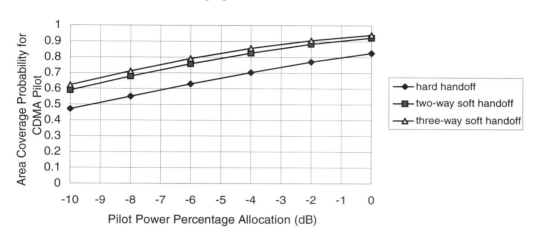

Figure 9–24 Pilot Power Percentage Allocation versus Area Coverage Probability for Hard and Soft Handoff with E_c/I_o Threshold −15 dB

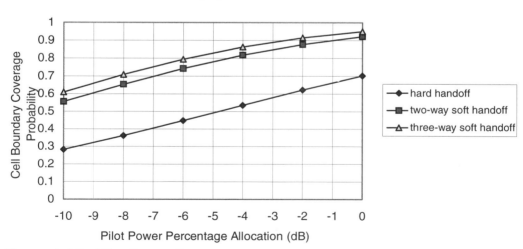

Figure 9–25 Pilot Power Percentage Allocation versus Cell Boundary Coverage Probability for Hard and Soft Handoff with E_c/I_o Threshold −15 dB

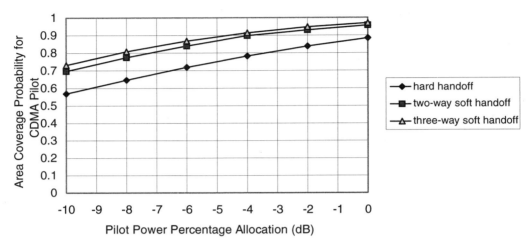

Figure 9-26 Pilot Power Percentage Allocation versus Area Coverage Probability for Hard and Soft Handoff with E_c/I_o Threshold −17.5 dB

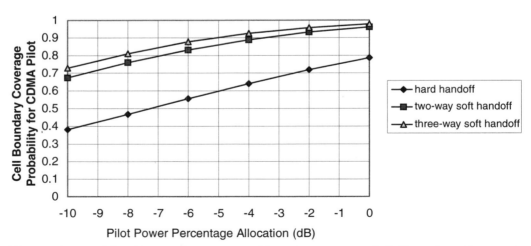

Figure 9-27 Pilot Power Percentage Allocation versus Cell Boundary Coverage Probability for Hard and Soft Handoff with E_c/I_o Threshold −17.5 dB

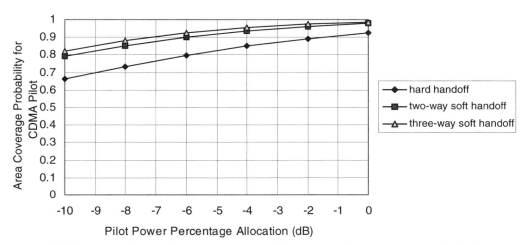

Figure 9–28 Pilot Power Percentage Allocation versus Area Coverage Probability for Hard and Soft Handoff with E_c/I_o Threshold –20 dB

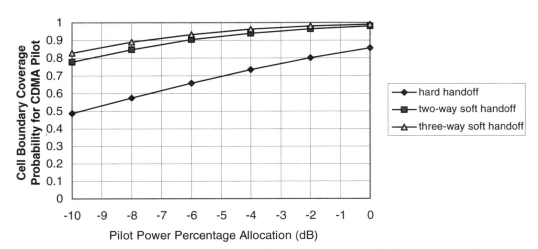

Figure 9–29 Pilot Power Percentage Allocation versus Cell Boundary Coverage Probability for Hard and Soft Handoff with E_c/I_o Threshold –20 dB

9.3 A Derivation of Coverage Probability for Pilot Channel

In this section, for hard and soft handoffs, we derive the analytical expressions for the probability that the ratio of the received pilot chip energy to spectral density of the thermal noise plus the interference is above the required threshold. By numerically evaluating the obtained expressions, trade-off among the system parameters (such as the CDMA pilot power allocation, the required E_c/I_o threshold, the use of hard or soft handoff, and the coverage probability) can be performed.

We use the following notations:

P_b = power per CDMA carrier at base station transmit filter input,
L_b^{Tx} = losses due to transmit filter, transmission lines, and cables at the transmit side of the base station,
$G_b^{Ant,Tx}$ = transmitter antenna gain of the base station,
$G_m^{Ant,Rx}$ = receiver antenna gain of the mobile station,
L_m^{Rx} = losses due to receive filter, transmission lines, and cables at the receiver side of the mobile station,
$L(r)$ = propagation path loss from base station to mobile station separated by distance r,
γ = propagation path loss exponent,
l_0 = path loss at intercept,
μ = fraction of base station power for pilot,
F = receiver noise figure of the mobile station,
N_{th} = thermal noise power spectral density,
R_{chip} = chip rate of pilot,
W = CDMA signal bandwidth, and
R_c = cell radius.

The pilot energy per chip that the mobile station receives is

$$E_{c,pilot} = \frac{\mu P_b}{R_{chip}} \times \frac{G_b^{Ant,Tx} G_m^{Ant,Rx}}{L(r_0^{(0)}) L_b^{Tx} L_m^{Rx}}. \tag{9.4}$$

When we assume that all other channel power contributes to the interference, the interference power spectral density generated from the same base station received by the mobile station is

$$I_{sc,o} = \frac{(1-\mu) P_b}{W} \times \frac{G_b^{Ant,Tx} G_m^{Ant,Rx}}{L(r_0^{(0)}) L_b^{Tx} L_m^{Rx}}. \tag{9.5}$$

All of the base stations are assumed to transmit at the same power level. Then the interference power spectral density generated from J adjacent base stations received by the mobile station is

$$I_{oc,o} = \frac{P_b}{W} \sum_{j=1}^{J} \frac{G_b^{Ant,Tx} G_m^{Ant,Rx}}{L(r_j^{(0)}) L_b^{Tx} L_m^{Rx}}. \qquad (9.6)$$

Note that in (9.5) and (9.6), $L(r_0^{(0)}) = l_0 \times r_0^{(0)}$ and $L(r_j^{(0)}) = l_0 \times r_j^{(0)}$, respectively.

Therefore, the ratio of received pilot energy per chip to power spectral densities of the thermal noise plus interference by the mobile station is

$$\frac{E_{c,\text{pilot}}}{I_o} = \frac{E_{c,\text{pilot}}}{FN_{\text{th}} + I_{sc,o} + I_{oc,o}}$$

$$= \frac{\dfrac{\mu P_b}{R_{chip}} \times \dfrac{G_b^{Ant,Tx} G_m^{Ant,Rx}}{L(r_0^{(0)}) L_b^{Tx} L_m^{Rx}}}{FN_{\text{th}} + \dfrac{(1-\mu)P_b}{W} \times \dfrac{G_b^{Ant,Tx} G_m^{Ant,Rx}}{L(r_0^{(0)}) L_b^{Tx} L_m^{Rx}} + \dfrac{P_b}{W} \sum_{j=1}^{J} \dfrac{G_b^{Ant,Tx} G_m^{Ant,Rx}}{L(r_j^{(0)}) L_b^{Tx} L_m^{Rx}}}$$

$$= \frac{\dfrac{W}{R_{chip}} \mu}{1 - \mu + L(r_0^{(0)}) \left[\dfrac{FN_{\text{th}} W L_b^{Tx} L_m^{Rx}}{P_b G_b^{Ant,Tx} G_m^{Ant,Rx}} + \sum_{j=1}^{J} \dfrac{1}{L(r_j^{(0)})} \right]}. \qquad (9.7)$$

Define the inverse of the CDMA carrier to thermal noise power ratio at the cell boundary as

$$\eta = \frac{FN_{\text{th}} W L_b^{Tx} L_m^{Rx} 10^{0.1[10\gamma \log_{10} R_c + 10 \log_{10} l_0]}}{P_b G_b^{Ant,Tx} G_m^{Ant,Rx}} \qquad (9.8)$$

Then, for $\dfrac{W}{R_{chip}} = 1$, we can rewrite (9.7) as

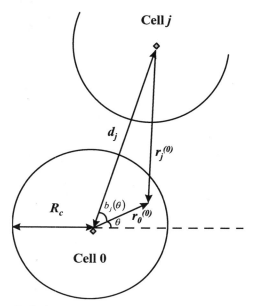

Figure 9–30 Distance Relationship Between Cell 0 and Cell j

$$\frac{E_{c,\text{pilot}}}{I_o} = \frac{\mu}{1 - \mu + L(r_0^{(0)})\left[\eta 10^{-0.1[10\gamma\log_{10}R_c + 10\log_{10}l_0]} + \sum_{j=1}^{J}\frac{1}{L(r_j^{(0)})}\right]}$$

$$= \frac{\mu}{\dfrac{\eta + \sum_{j=1}^{J}\dfrac{10^{0.1[10\gamma\log_{10}R_c + 10\log_{10}l_0]}}{L(r_j^{(0)})}}{\dfrac{10^{0.1[10\gamma\log_{10}R_c + 10\log_{10}l_0]}}{L(r_0^{(0)})}} - \mu} \qquad (9.9)$$

It should be noted that the index $j = 0$ represents the in-cell lognormal shadowing and all other indices $j \neq 0$ represent the outer-cell lognormal shadowing.

Next, we define

$$Y = 10\log\sum_{j=0}^{J}\frac{10^{0.1[10\gamma\log_{10}R_c + 10\log_{10}l_0]}}{L(r_j^{(0)})} = 10\log\sum_{j=0}^{J}10^{0.1Z_j}, \qquad (9.10)$$

where, for $j = 0, 1, 2, \cdots, J$, Z_j is given by

$$Z_j = -10\log_{10} L(r_j^{(0)}) + 10\gamma \log_{10} R_c + 10\log_{10} l_0 = -10\gamma \log_{10} \frac{r_j^{(0)}}{R_c}. \qquad (9.11)$$

Then, Z_j is a Gaussian random variable with mean and variance given as

$$E\{Z_0\} = 10\log_{10}\left(\frac{r_0^{(0)}}{R_c}\right)^{-\gamma} \qquad (9.12)$$

$$E\{Z_j\} = 10\log_{10}\left(\sqrt{\left(\frac{r_0^{(0)}}{R_c}\right)^2 + a_j - 2\sqrt{a_j}\frac{r_0^{(0)}}{R_c}\cos b_j(\theta)}\right)^{-\gamma}, \quad j = 1, 2, \cdots, J \qquad (9.13)$$

$$\text{Var}\{Z_j\} = \sigma^2, \quad j = 0, 1, 2, \cdots, J \qquad (9.14)$$

In (9.13), $a_j = (d_j/R_c)^2$ and d_j is the distance between cell 0 and cell j, as shown in Figure 9–30.

Since $10^{0.1Y}$ is the sum of lognormal random variables, and the probability density function $P_Y(y)$ can be assumed to be Gaussian, its mean ε_Y and variance σ_Y^2 can be calculated from the means and variances of the random variables Z_js for $j = 0, 1, \cdots, J$ (Reference [9.1]).

The strongest pilot is obtained among $(J + 1)$ base stations. That is,

$$\begin{aligned} X \equiv Z_0 &= -10\log L(r_0^{(0)}) + 10\gamma \log_{10} R_c + 10\log_{10} l_0 \\ &= \max_j\{-10\log L(r_j^{(0)}) + 10\gamma \log_{10} R_c + 10\log_{10} l_0, \quad j = 0, 1, 2, \cdots, J\} \\ &= \max_j\{Z_j, \quad j = 0, 1, 2, \cdots, J\}. \end{aligned} \qquad (9.15)$$

Now that Z_js are Gaussian random variables with mean and variance given by (9.12), (9.13), and (9.14), respectively, the probability density function of X is

$$p_X(x) = \sum_{j=1}^{J} \frac{1}{\sqrt{2\pi\sigma^2}} \exp\left[-\frac{(x-\varepsilon_j)^2}{2\sigma^2}\right] \prod_{\substack{i=1 \\ i \neq j}}^{J}\left[1 - Q\left(\frac{x-\varepsilon_i}{\sigma}\right)\right], \qquad (9.16)$$

where ε_j denotes the mean of (9.13) and $Q(\cdot)$ is the Q function described in Section 7.1.2.

Thus, the ratio of received pilot energy per chip to spectral densities of the thermal noise plus interference by the mobile station (that is, Equation (9.9)) can be written as

$$\frac{E_{c,\text{pilot}}}{I_o} = \frac{\mu}{\dfrac{\eta + 10^{0.1Y}}{10^{0.1X}} - \mu}. \tag{9.17}$$

Let δ be the required $E_{c,\text{pilot}}/I_o$ threshold to achieve certain coverage probability. The coverage probability at a location $(r_0^{(0)}, \theta)$ is thus

$$\Pr\left\{\frac{E_{c,\text{pilot}}}{I_o} \geq \delta\right\} = \Pr\left\{10^{0.1Y} \leq \left(1 + \frac{1}{\delta}\right)\mu 10^{0.1X} - \eta\right\}$$

$$= \Pr\left\{Y \leq 10\log_{10}\left[\left(1+\frac{1}{\delta}\right)\mu 10^{0.1X} - \eta\right], X > 10\log_{10}\frac{\eta}{\left(1+\frac{1}{\delta}\right)\mu}\right\}$$

$$= \int_{10\log_{10}\frac{\eta}{\left(1+\frac{1}{\delta}\right)\mu}}^{\infty} \left[1 - \int_{10\log_{10}\left[\left(1+\frac{1}{\delta}\right)\mu 10^{0.1x} - \eta\right]}^{\infty} p_Y(y)dy\right] p_X(x)dx$$

$$= \int_{10\log_{10}\frac{\eta}{\left(1+\frac{1}{\delta}\right)\mu}}^{\infty} \left\{1 - Q\left(\frac{10\log_{10}\eta + 10\log_{10}\left[\left(1+\frac{1}{\delta}\right)\frac{\mu}{\eta}10^{0.1x} - 1\right] - \varepsilon_Y}{\sigma_Y}\right)\right\} p_X(x)dx. \tag{9.18}$$

9.4 Reference

[9.1] S. C. Schwartz and Y. S. Yeh. On the Distribution Function and Moments of Power Sums With Log-Normal Components, *The Bell System Technical Journal*, vol. 61, pp. 1441-1462, September 1982.

CHAPTER 10

Traffic Engineering

10.1 Introduction

In a CDMA system, the channel element (CE) performs the baseband spread-spectrum signal processing for a given channel (pilot, sync, paging, or traffic channel). Erlang capacity is determined not only by the maximum number of CEs available for traffic channels, but also by the maximum number of simultaneous active users.

At a three-sector CDMA cell, it would be wasteful to provision CEs per sector based on the per-sector traffic loads because the trunking efficiency gained by pooling the CEs would be lost. When all the CEs are pooled, any CE can be assigned to any user in the cell, regardless of sector. As another factor affecting provisioning, the CDMA system must impose a limit on the number of simultaneous users in a sector to control the interference between users having the same pilot. Therefore, it would be inappropriate to provision CEs for base stations based just on the total load because such an approach neglects the per-sector limits. The optimum procedure must account for blocking that occurs when users saturate a sector and when all CEs at the base station are busy. To accommodate a soft handoff requires the allocation of more CEs. We recommend an additional thirty-five percent initially although this may need to be adjusted at individual base stations based on the actual percentage of mobile stations in soft handoff.

Section 10.2 presents an algorithm that computes blocking probability for a CDMA base station based on the individual traffic loads of the sectors and the total number of CEs. The optimum is the smallest number of CEs that meets the blocking objective. To illustrate the use of this algorithm, Section 10.3 presents extensive numerical results (in figures and tables) of the calculated Erlang capacity per sector. These numerical results are calculated for (a) one CDMA carrier; (b) users per sector per carrier limited from ten users to fifteen users; and (c) two percent blocking objective.

It should be noted that the required number of traffic CEs reported in this chapter does not include the additional CEs for soft handoff and other overhead channels for pilot, synchronization, and paging.

10.2 Analysis

In this section we derive an algorithm for computing blocking probability of sectored cells in a CDMA system. A MATLAB program that implements the algorithm has been written.

As background, Poisson arrivals and exponentially distributed holding times characterize the traffic impinging on a cell. If λ denotes the arrival rate of calls in a region and T the average holding time, then the traffic load in Erlangs is

$$a = \lambda T.$$

The traffic load for the three sectors will be denoted by (a,b,c). We assume that blocked calls are cleared and the maximum number of users in a sector is M, irrespective of loading. The probability of having exactly n users in sector A is depicted in Reference [10.1] and given by

$$P_A(n) = \frac{\frac{a^n}{n!}}{\sum_{m=0}^{M} \frac{a^m}{m!}}, \quad 0 \leq n \leq M.$$

Note that $P_A(M)$ is the blocking probability according to Erlang B. Solving the associated birth-death equations and imposing the condition that the sum of the probabilities equals one finds these occupation probabilities. The same equation applies for sectors B and C.

The algorithm is an iterative one and hence we start by discussing the two-sector case.

10.2.1 Two-Sector Case

If we denote two sectors as A and B and the column vectors of their respective marginal probability as $\mathbf{P_A}$ and $\mathbf{P_B}$ then

Analysis

$$P_A(i) = \sum_{j=0}^{M} P_{A|B}(i|j) P_B(j) \quad \text{for } 0 \le i \le M, \tag{10.1}$$

where $P_A(i)$ and $P_B(i)$ are entries of $\mathbf{P_A}$ and $\mathbf{P_B}$, respectively, and $P_{A|B}(i|j)$ is the conditional probability. Note that $P_{A|B}(i|j) = 0$ when $i + j > N$, N being the number of CEs in a cell. The above equation can be expressed in a vector form as

$$\begin{pmatrix} P_A(0) \\ P_A(1) \\ \vdots \\ P_A(M) \end{pmatrix} = \begin{pmatrix} P_{A|B}(0|0) & P_{A|B}(0|1) & \cdots & P_{A|B}(0|M) \\ P_{A|B}(1|0) & P_{A|B}(1|1) & \cdots & P_{A|B}(1|M) \\ \vdots & \vdots & \ddots & \vdots \\ P_{A|B}(M|0) & P_{A|B}(M|1) & \cdots & 0 \end{pmatrix} \begin{pmatrix} P_B(0) \\ P_B(1) \\ \vdots \\ P_B(M) \end{pmatrix}$$

or

$$\mathbf{P_A} = \mathbf{P_{A|B}} \mathbf{P_B}. \tag{10.2}$$

The j-th column of the conditional matrix $\mathbf{P_{A|B}}$ is the state probability vector of sector A, given that sector B has j users. In this case, traffic impinging on sector A will have $N - j$ channel elements available and the state probability of sector A will be the same as that of an omni-cell having $N - j$ CEs, that is,

$$\mathbf{p_{A|B}}(\cdot|j) = \begin{cases} \dfrac{1}{\sum_{k=0}^{M} \dfrac{a^k}{k!}} \times \begin{pmatrix} 1 \\ a \\ \vdots \\ \dfrac{a^M}{M!} \end{pmatrix}, & \text{for } 0 \le j \le N - M \\[2em] \dfrac{1}{\sum_{k=0}^{N-j} \dfrac{a^k}{k!}} \times \begin{pmatrix} 1 \\ a \\ \vdots \\ \dfrac{a^{N-j}}{(N-j)!} \\ 0 \\ \vdots \\ 0 \end{pmatrix}, & \text{for } N - M + 1 \le j \le M \end{cases} \tag{10.3}$$

10.2.1.1 Equal Erlang Load

If the two sectors A and B are subjected to the same Erlang load, then $\mathbf{P_A} = \mathbf{P_B}$. Therefore (10.2) becomes

$$\mathbf{P_A} = \mathbf{P_{A|B}} \mathbf{P_A}. \tag{10.4}$$

Equation (10.4) is a special case of the eigenvalue problem (Reference [10.2]); that is, $\mathbf{P_A}$ is an eigenvector of $\mathbf{P_{A|B}}$ with the eigenvalue of one. Therefore, given the conditional matrix $\mathbf{P_{A|B}}$ of (10.3), one can find the marginal probability vector $\mathbf{P_A}$.

In addition, the joint probability matrix $\mathbf{P_{AB}}$ can be obtained from $\mathbf{P_{A|B}}$ and $\mathbf{P_A}$ as

$$\mathbf{P_{AB}} = \begin{pmatrix} P_{AB}(0,0) & P_{AB}(0,1) & \cdots & P_{AB}(0,M) \\ P_{AB}(1,0) & P_{AB}(1,1) & \cdots & P_{AB}(1,M) \\ \vdots & \vdots & \ddots & \vdots \\ P_{AB}(M,0) & P_{AB}(M,1) & \cdots & 0 \end{pmatrix}$$

$$= \left(P_A(0) \begin{pmatrix} P_{A|B}(0|0) \\ P_{A|B}(1|0) \\ \vdots \\ P_{A|B}(M|0) \end{pmatrix}, P_A(1) \begin{pmatrix} P_{A|B}(0|1) \\ P_{A|B}(1|1) \\ \vdots \\ P_{A|B}(M|1) \end{pmatrix}, \cdots, P_A(M) \begin{pmatrix} P_{A|B}(0|M) \\ P_{A|B}(1|M) \\ \vdots \\ 0 \end{pmatrix} \right). \tag{10.5}$$

Once the joint probability matrix $\mathbf{P_{AB}}$ is determined, the blocking probability for sector A is given by

$$P_{\text{block}} = \sum_{j=0}^{N-M} P_{AB}(M,j) + \sum_{j=N-M+1}^{M} P_{AB}(N-j,j). \tag{10.6}$$

We assume here that $N < 2M$, which implies that when in the blocking state, all CEs are occupied.

10.2.1.2 Unequal Erlang Load

In this case, (10.4) does not hold. However, following (10.2), one can write

$$\mathbf{P_B} = \mathbf{P_{B|A}} \mathbf{P_A}. \tag{10.7}$$

Substituting (10.7) for $\mathbf{P_B}$ in (10.2), one can rewrite (10.2) as follows

$$\mathbf{P_A} = \mathbf{P_{A|B}} \mathbf{P_{B|A}} \mathbf{P_A}. \tag{10.8}$$

In (10.8), the marginal probability $\mathbf{P_A}$ is the eigenvector of the product of the two conditional probability matrices. Note that the matrix $\mathbf{P_{B|A}}$ is not the inverse of $\mathbf{P_{A|B}}$.[1] In addition, the conditional probability matrices would not be of full rank.

Note that the equal Erlang load case can be considered as a special case of (10.8). If the sectors have equal Erlang load (that is, $a = b$), then $\mathbf{P_{A|B}} = \mathbf{P_{B|A}}$. Therefore (10.8) becomes

$$\mathbf{P_A} = \mathbf{P_{A|B}^2 P_A},$$

and $\mathbf{P_A}$ is the eigenvector of $\mathbf{P_{A|B}^2}$ whose eigenvalue is one. Because $\mathbf{P_A}$ is the eigenvector of $\mathbf{P_{A|B}}$ with an eigenvalue of one as discussed with (10.4), $\mathbf{P_A}$ is also the eigenvector of $\mathbf{P_{A|B}^2}$ with an eigenvalue of one.[2]

10.2.2 Three-Sector Case

We denote the three sectors by A, B and C and the respective marginal probability by the column vectors $\mathbf{P_A}$, $\mathbf{P_B}$, and $\mathbf{P_C}$. We will establish an eigenvector equation for the three-sector case following the procedure used for the two-sector case. We will also use the results of the two-sector cell derived in the previous section.

Assuming the different sectors are equally loaded (that is, $a = b = c$), one can write

$$\mathbf{P_A} = \mathbf{P_{A|B+C} P_{B+C}}$$

and

$$\mathbf{P_{B+C}} = \mathbf{P_{B+C|A} P_A}.$$

It then follows that

$$\mathbf{P_A} = \mathbf{P_{A|B+C} P_{B+C|A} P_A}, \tag{10.9}$$

or the marginal probability vector $\mathbf{P_A}$ is the eigenvector of the matrix given by $\mathbf{P_{A|B+C} P_{B+C|A}}$ whose eigenvalue is one. In (10.9), the matrices $\mathbf{P_{A|B+C}}$ and $\mathbf{P_{B+C|A}}$ are determined as explained below.

The conditional matrix $\mathbf{P_{A|B+C}}$ is the same as the conditional matrix for the two-sector case. The columns of $\mathbf{P_{A|B+C}}$ are given by (10.3), where the index j is now the sum of the number of users in sectors B and C. The conditional matrix $\mathbf{P_{B+C|A}}$ can be

---

1. In fact, $\mathbf{P_{B|A}}$ will have the same structure as $\mathbf{P_{A|B}}$ given by (10.3). For $\mathbf{P_{B|A}}$, traffic load b replaces a in (10.3).
2. It can be shown that every eigenvector of a matrix $\mathbf{A}$ is also an eigenvector of $\mathbf{A}^2$; that is, if $\mathbf{Ax} = \lambda \mathbf{x}$, then $\mathbf{A}^2 \mathbf{x} = \lambda^2 \mathbf{x}$ (λ is the eigenvalue, not the call arrival rate).

determined from the joint probability matrix of the two-sector case. The columns of $\mathbf{P_{B+C|A}}$ are given by

$$\mathbf{P_{B+C|A}}(\cdot\,|\,j) = \begin{pmatrix} P_{BC}(0,0) \\ P_{BC}(0,1) + P_{BC}(1,0) \\ \vdots \\ \sum_{k=0}^{\min(i,N-j)} P_{BC}(k,i-k) \\ \vdots \\ \sum_{k=0}^{\min(M,N-j)} P_{BC}(k,M-k) \end{pmatrix},$$

where $\mathbf{P_{BC}}$ is the same as $\mathbf{P_{AB}}$ given in (10.5) with $N-j$ channel elements available.

Solving (10.9) for $\mathbf{P_A}$, one can then determine the joint probability matrix $\mathbf{P_{B+C,A}}$ as

$$\mathbf{P_{B+C,A}} = \begin{pmatrix} P_A(0) \cdot \begin{pmatrix} P_{B+C|A}(0|0) \\ P_{B+C|A}(1|0) \\ \vdots \\ P_{B+C|A}(2M|0) \end{pmatrix}, P_A(1) \cdot \begin{pmatrix} P_{B+C|A}(0|1) \\ P_{B+C|A}(1|1) \\ \vdots \\ P_{B+C|A}(2M|1) \end{pmatrix}, \cdots, P_A(M) \cdot \begin{pmatrix} P_{B+C|A}(0|M) \\ P_{B+C|A}(1|M) \\ \vdots \\ 0 \end{pmatrix} \end{pmatrix}.$$

From the joint probability matrix $\mathbf{P_{B+C,A}}$, the probability of blocking is given by

$$P_{\text{block}} = \sum_{j=0}^{N-M} P_{B+C,A}(j,M) + \sum_{j=N-M+1}^{M} P_{B+C,A}(j,N-j). \qquad (10.10)$$

We assume here that $N < 3M$, which implies that when in the blocking state, all CEs are occupied.

10.2.3 K-Sector Case

In the previous section, we formulated the problem of finding the marginal probability of a given sector in a three-sector base station as that of finding the eigenvector with an eigenvalue of one of a matrix. This matrix was the product of two conditional probability matrices. The first was the conditional probability matrix of one sector given the sum of the number of CEs used by the other sectors, which can be derived from the traditional Erlang B model. The second was the conditional probability matrix on the number of CEs used by two sectors given the number occupied by the sector

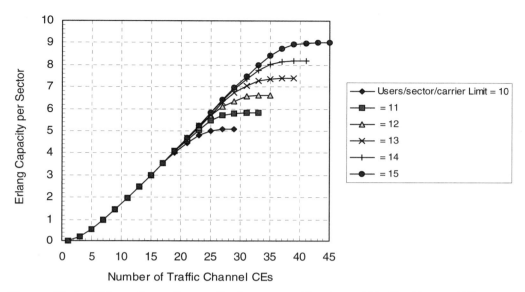

Figure 10–1 Erlang Capacity per Sector versus Number of Traffic Channel Elements in Three-Sector Base Station for One CDMA Carrier per Sector with Equal Sector Loads and Two Percent Blocking Probability

under consideration, which is derived from the results of the two-sector case in Section 10.2.1.

Similarly, the marginal probability of a sector in a configuration where K sectors use CEs from the same pool is the eigenvector with an eigenvalue of one of a product of two matrices. The first is the conditional probability matrix of the sector under consideration given the sum of the number of CEs used by the other $K - 1$ sectors, which can be derived from the traditional Erlang B model. The second is the conditional probability matrix on the number of CEs used by $K - 1$ sectors given the number occupied by the sector under consideration, which is derived from the $K - 1$ sector case.

By denoting the sectors by $A_1, A_2, ..., A_K$, we obtain the marginal probability vector of A_1, P_{A_1}, by solving the following eigenvector equation

$$P_{A_1} = P_{A_1 | A_2 + ... + A_K} \cdot P_{A_2 + ... A_K | A_1} \cdot P_{A_1},$$

where $P_{A_1 | A_2 + ... + A_K}$ is the conditional probability matrix of the sector under consideration given the sum of the number of CEs used by the other $K - 1$ sectors and $P_{A_2 + ... A_K | A_1}$ is the conditional probability matrix on the number of CEs used by $K - 1$ sectors given the number occupied by the sector under consideration. Determination of

Table 10–1 Erlang Capacity per Sector versus Number of Traffic Channel Elements in a Three-Sector Base Station for One CDMA Carrier per Sector with Equal Sector Loads and Two Percent Blocking Probability

Number of CEs In Three-Sector BS	Users per Sector per Carrier Limit					
	10	11	12	13	14	15
1	0.01	0.01	0.01	0.01	0.01	0.01
2	0.07	0.07	0.07	0.07	0.07	0.07
3	0.2	0.2	0.2	0.2	0.2	0.2
4	0.36	0.36	0.36	0.36	0.36	0.36
5	0.55	0.55	0.55	0.55	0.55	0.55
6	0.76	0.76	0.76	0.76	0.76	0.76
7	0.98	0.98	0.98	0.98	0.98	0.98
8	1.21	1.21	1.21	1.21	1.21	1.21
9	1.45	1.45	1.45	1.45	1.45	1.45
10	1.69	1.69	1.69	1.69	1.69	1.69
11	1.95	1.95	1.95	1.95	1.95	1.95
12	2.2	2.2	2.2	2.2	2.2	2.2
13	2.47	2.47	2.47	2.47	2.47	2.47
14	2.73	2.73	2.73	2.73	2.73	2.73
15	2.99	3	3	3	3	3
16	3.27	3.28	3.28	3.28	3.28	3.28
17	3.53	3.54	3.55	3.55	3.55	3.55
18	3.72	8.82	3.83	3.83	3.83	3.83
19	4	4.09	4.1	4.11	4.11	4.11
20	4.27	4.35	4.38	4.39	4.39	4.39
21	4.46	4.57	4.66	4.67	4.68	4.68
22	4.46	4.86	4.94	4.96	4.97	4.97
23	4.81	5.03	5.14	5.23	5.24	5.25
24	4.88	5.29	5.43	5.51	5.53	5.54
25	5.01	5.46	5.69	5.72	5.81	5.82
26	5.03	5.58	5.91	6.02	6.1	6.12
27	5.07	5.71	6.09	6.3	6.38	6.41
28	5.07	5.78	6.27	6.5	6.51	6.7
29	5.08	5.8	6.35	6.74	6.9	6.97
30	5.08	5.83	6.49	6.93	7.15	7.2
31		5.84	6.56	7.06	7.38	7.5
32		5.82	6.59	7.21	7.58	7.77
33		5.84	6.61	7.29	7.76	7.99
34			6.61	7.35	7.9	8.18
35			6.61	7.38	8.02	8.42
36			6.58	7.39	8.09	8.56
37				7.4	8.15	8.74
38				7.4	8.18	8.84
39				7.39	8.2	8.92
40					8.2	8.96
41					8.2	8.99
42						9.01
43						9.01
44						9.01
45						9.01

the marginal probability allows derivation of the state probabilities, which can be used in determining the blocking probability.

The K-sector method described could be useful in assessing blocking characteristics where CEs are shared across more than three sectors (for example, a six-sector, single-carrier cell) or across more than one carrier (for example, a three-sector, two-carrier cell).

10.3 Numerical Results

For three-sector CDMA cells, assuming that the sectors are equally loaded, we provide calculated Erlang capacity per sector. Table 10–1 and Figure 10–1 show this result for one CDMA carrier per sector with a two percent blocking objective and different limits on the number of users per sector per carrier.

10.4 References

[10.1] D. Bertsekas and R. Gallager. *Data Networks*, Prentice-Hall, Inc., Englewood Cliffs, New Jersey, 1987.

[10.2] T. Kailath. *Linear Systems*, Prentice-Hall, Inc., Englewood Cliffs, New Jersey, 1980.

CHAPTER 11

Antennas

11.1 Introduction

The antenna, as a subsystem including antenna and feed, transmits or receives radio waves. Its basic function is to couple electromagnetic (EM) energy between free space and a guiding device such as a transmission line, coaxial cable, or waveguide. In wireless communication systems, the antenna is one of the most critical components; it can either enhance or constrain system performance.

For example, antenna radiation pattern plays an important role in improving the capacity of mobile communications systems. Sectored cells use directional antennas, and these antennas serve only a particular direction in a sectored cell, thus reducing co-channel interference and increasing capacity, which is more pronounced in CDMA mobile communication systems.

Antenna diversity is another important issue in wireless communication systems. Multipath propagation due to many paths (reflection, diffraction, and scattering) causes fading that results in rapid variations in the received signal strength. With the antenna diversity at the base station or mobile, either spatially separated antennas (space diversity) or orthogonally polarized antennas (polarization diversity) receive signals to reduce the severity of fading and to provide significant link improvement of the reception.

11.2 Antenna Concepts

Some important antenna concepts, such as antenna radiation pattern, directivity, gain, efficiency, and polarization, illustrate the characteristics of the type of antennas used in modern wireless communication systems.

Antenna Radiation Pattern: This is a graphical representation of the radiation properties of an antenna as a function of space coordinates. Radiation properties include radiation intensity, field strength, phase, and polarization. In most cases, the radiation pattern, which is represented as a function of directional coordinates, is determined in the far-field region generally defined by $r \geq 2d^2/\lambda$, where r is a distance from antenna, λ is the wavelength, and d is a maximum overall dimension of the antenna. The antenna radiation pattern includes the main lobe and side lobe. The main lobe is the radiation lobe containing the direction of maximum radiation. The side lobe is a radiation lobe in any direction other than that of the main lobe. The amplitude level of a side lobe relative to the main lobe is referred to as the *side-lobe level*.

Wireless communication uses two kinds of antenna pattern: the omnidirectional antenna has an essentially nondirectional pattern in azimuth and a directional pattern in elevation and the directional antenna has the directional pattern in both azimuth and elevation.

Antenna Radiation Beamwidth: This is the angular separation between two directions in which the radiation intensity is identical, with no other intermediate points of the same value. When the intensity is half of the maximum, it is referred to as *half-power beamwidth*.

Antenna Directivity **D**: This is the ratio of the maximum radiation intensity in a given direction (usually zero degrees or foresight) from an antenna to the radiation intensity averaged over all directions. In mathematical form, directivity can be written as

$$D = \frac{4\pi U_{max}}{P} = \frac{4\pi}{\int_0^{2\pi}\int_0^{\pi} S(\theta,\phi)\sin\theta d\theta d\phi}. \qquad (11.1)$$

Antenna Gain $G(\theta,\varphi)$: This is the ratio of the radiation intensity in a given direction to the radiation intensity that would be obtained if the power accepted by the antenna were radiated isotropically. The gain of an antenna is related to the directivity and also takes into account antenna efficiency. The antenna power gain is defined as 4π times the ratio of the radiation intensity in that direction to the net power accepted by the antenna from a connected transmitter. The antenna gain is related to the antenna directivity by the antenna efficiency, η. That is,

$$G(\theta,\phi) = \eta D(\theta,\phi). \qquad (11.2)$$

The D in (11.2) is the directivity of (11.1) measured in the *specific* direction (θ,φ). The antenna efficiency is used to take into account losses at the input terminals and within the structure of the antenna. The antenna gain is expressed with respect to the ideal isotropic radiation pattern, in which case it is measured in dBi. Wireless mobile communications systems usually use high directivity, high gain antennas because the signals may go through severe fading.

Antenna Polarization: The polarization of an antenna is the polarization of the wave radiated by the antenna in a given direction when the antenna is excited. In general, the polarization of an antenna is classified as linear, circular, or elliptical. In linear polarization, if the electric force lines are parallel to the surface of the earth, the wave is called a horizontally polarized wave; similarly, if the electric force line is perpendicular to the surface of the earth, the wave is called a vertically polarized wave.

Input impedance: This is the ratio of the voltage to current at a pair of antenna input terminals. The value of the antenna input impedance is dependent on the shape of the antenna, wavelength, and surrounding objects (including other nearby antennas). Antenna input impedance is important to the transfer of power from a transmitter to an antenna or from an antenna to a receiver. The maximum power can be transferred to/from the antenna if the impedance between the antenna and transmission line is well matched.

References [11.1]-[11.3] discuss the characteristics of some typical antennas and antenna arrays.

11.3 Antenna System with Interference and Cell Coverage

Antenna pattern, gain, height, and orientation (for example, antenna tilting) all affect the wireless system design. The antenna pattern can be omnidirectional or directional in both the vertical and the horizon planes. Antenna gain compensates for losses in the transmitted and received signal power. The antenna height of a cell site can affect the area and shape of the cell coverage. The antenna downtilt can reduce the interference to the neighboring cells and enhance the coverage for the weak spots within the cell.

11.3.1 Directional Antenna and Sectorization Gain

Since CDMA systems typically employ universal frequency reuse, co-channel interference will occur. The directional antennas in a sectored cell can reduce co-channel interference and enhance system capacity. In a base station of 120-degree cell sectors, the antenna system includes three directional transmit antennas and six directional

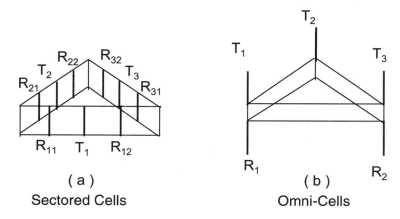

Figure 11–1 Base Station Antenna Configurations

receive antennas. Figure 11–1(a) shows an example of this kind of antenna configuration. There are one transmitter and two (diversity) receivers in each sector. The gain for each antenna, for example, is 15 to 19 dBi with about a ninety-degree half-power radiation pattern at horizontal plane and five-and-a-half-degree to seven-degree beamwidth at vertical plane. Figure 11–1(b) illustrates an example of the antenna system of an omni-cell that includes three omnidirectional transmit antennas (possibly for three carriers) and two omnidirectional receive antennas. Omnidirectional antennas have an 8 to 12 dBi gain, 360-degree horizontal beamwidth, and about a seven-degree vertical beamwidth.

In three-sector cell sites, an ideal 120-degree antenna pattern, as shown in Figure 11–2 (a), will receive signals from only one-third of the cell, reducing the interference by two-thirds; thus the sector gain is three, resulting in the CDMA channel capacity gain of a factor of three. However, the radiation pattern of actual antennas (see Figure 11–2 (b)) would be different from the ideal one shown in Figure 11–2 (a).

In reality, there is always some overlap of the sector antenna patterns, so interference would not be reduced exactly by a factor of three. The sectorization gain can be expressed as

$$\chi = \text{Number\_of\_Sectors} \times \left(\frac{1 + \beta_{omni}}{1 + \beta_{sector}} \right), \tag{11.3}$$

where β_{omni} is the interference factor for omnidirectional cells and β_{sector} is the interference factor for sectored cells.

Antenna System with Interference and Cell Coverage

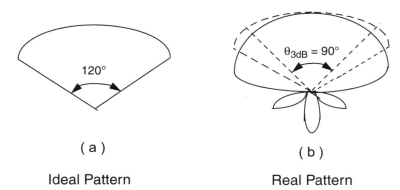

Figure 11–2 Directional Antenna Pattern in a 120-Degree Sector

As shown in Table 8–1, $\beta_{omni} = 0.6$ and $\beta_{sector} = 0.85$ for three-sector cells. In this case, the sectorization gain is

$$\chi = 3 \times \left[\frac{1+0.6}{1+0.85}\right] = 2.6.$$

We define the sectorization efficiency, κ, as

$$\kappa = \frac{\chi}{\text{Number\_of\_Sectors}}. \qquad (11.4)$$

For example, the sectorization efficiency of the above three-sector cell is 86.5 percent.

The following equation gives the single-sector RF channel capacity of a sectored cell in terms of an omnidirectional cell capacity and the sectorization efficiency:

$$M_{sector} = M_{omni} \times \kappa, \qquad (11.5)$$

where M_{sector} is a single-sector RF channel capacity of a sectored cell and M_{omni} is an omnidirectional cell capacity.

11.3.2 Coverage with Antenna Height and Gain

The radio propagation shown in Figure 11–3 is based on geometric optics and geometrical theory of diffraction (GTD) and considers the direct path, a ground reflected propagation path, and a diffracted propagation path from the building edge between the transmitter and receiver.

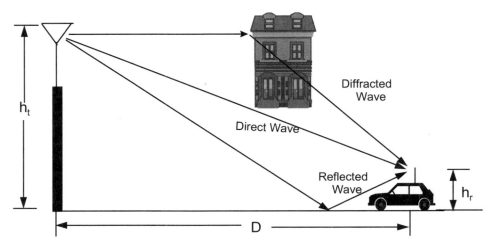

Figure 11–3 Radio Propagation from Transmitter to Receiver

The coverage distance D from the transmitter to receiver can be expressed as

$$D \approx \left(\frac{P_t}{P_r}\right)^{1/4} (h_r h_t)^{1/2} [G_r G_t]^{1/4} (L_a)^{1/4}, \tag{11.6}$$

where

P_r is the received power,
P_t is the transmitted power,
h_r is the height of the receiver,
h_t is the height of the transmitter,
G_r is the receiver antenna gain,
G_t is the transmitter antenna gain, and
L_a is a correction factor that includes the diffracted path loss and others.

As seen from (11.6), the coverage is proportional to the height of the antenna and antenna gain. In cellular and PCS systems, the base station antenna heights range from twenty to one-hundred meters; antenna heights are dependent on the different environments. For example, the antenna height is about 30 m in an urban area, about 50 m in a suburban area, and about 80 m in a rural area. Also, the requirement of the antenna gain depends on the environment. In rural highway sites, the higher antenna gain (for example, 19 dBi) may be used to extend coverage along the highway. However, in

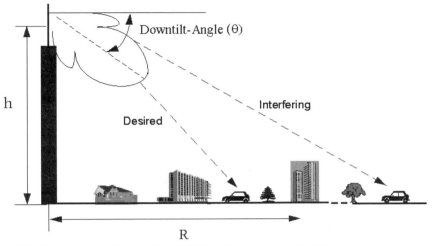

Figure 11-4 Antenna Beam Downtilt for Illuminating the Service Area

urban sites, the lower antenna gain (for example, 15 dBi) has to be used to reduce the interference.

11.3.3 Reducing Interference Using Antenna Downtilt

The principal idea of the antenna downtilt technique is to tilt the main beam to a certain angle to suppress the power level toward the reuse cell site and to reduce co-channel interference. This downtilt can be accomplished by mechanical or electrical means. Tilting the antenna vertical beam pattern reduces the interference to other cell sites because the received field intensity in these cell sites is weak. This is an advantage from the viewpoint of system design. Figure 11-4 shows a model for typical antenna beam pattern downtilt.

In PCS and cellular communications, the antenna downtilt angle, θ, is a function of the antenna height, coverage radius, and antenna vertical beamwidth. In general, when the coverage radius of a service area is set to a specified value, the higher the antenna, the larger the downtilt angle, and the larger the reduction in co-channel interference. On the other hand, when the height of the antenna in the base station is set, the smaller the coverage radius, the larger the downtilt angle. For different engineering considerations, two formulas of antenna downtilt can be expressed as

$$\theta = \tan^{-1}\left(\frac{h}{2R}\right) + \frac{HPBW_{ver}}{2} \quad (11.7)$$

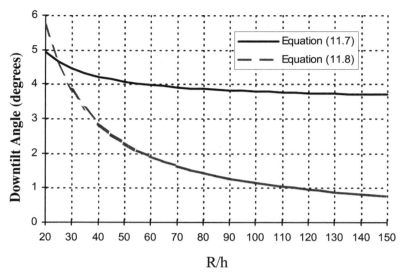

Figure 11–5 The Angle of Antenna Downtilt as a Function of R/h for the Vertical Beamwidth of Seven Degrees.

$$\theta = \pi - 2\tan^{-1}\left(\frac{R}{h}\right), \qquad (11.8)$$

where $HPBW_{ver}$ denotes the antenna vertical half-power beamwidth. Equation (11.7) represents the angle of antenna downtilt required to reduce the interference at the base of the neighbor cell ($D = 2R$) by 3 dB. Equation (11.8) represents the angle of antenna downtilt that would be needed to preserve the coverage in the fringe of the cell ($D = R$).

Figure 11–5 shows the angle of antenna downtilt as a function of R/h (coverage radius/antenna height) for the vertical beamwidth of seven degrees. In Figure 11–5, the solid line is the first downtilt formula (11.7) and the dashed line is the second formula (11.8). As the plot shows, the first formula for the downtilt angle prediction shows less dependence on the R/h ratio. For a small R/h (small cell radius and/or high antenna), the second formula predicts a larger downtilt angle, while for a large R/h (large cell radius and/or low antenna), the second formula predicts a smaller downtilt angle.

Table 11–1 shows the typical antenna downtilt angle for different morphology classes in PCS communication systems.

Table 11–1 Antenna Downtilt Angle in PCS Communication Systems

Morphology Class	Antenna Height (m)	Coverage Radius (Km)	Vertical Beamwidth (degrees)	Downtilt Using Eq. (11.7)	Downtilt Using Eq. (11.8)	Average Downtilt (degrees)
Dense Urban	30	0.8	7	4.57	4.30	4.43
Light Urban	25	1.5	7	3.98	1.91	2.94
Suburban	20	3.0	7	3.69	0.76	2.23
Rural	50	5.0	7	3.79	0.57	2.18

11.4 Diversity Antenna Systems

Since signal fading in the wireless radio environment causes severe reception problems, diversity antenna techniques are used to reduce the fading effects. Usually the diversity is applied at the base station. Space diversity and polarization diversity are two commonly used techniques for PCS and cellular systems.

11.4.1 Space Diversity Antenna

One can achieve space diversity using multiple antennas separated by finite distance. The spatial separation between the multiple antennas is chosen so that the diversity branches experience uncorrelated or nearly-uncorrelated fading. A typical sectored base station configured with space diversity consists of three antennas per sector, one transmit and two space-separated receive antennas, as illustrated in Figure 11–6.

The cross-correlation between two received signals at a base station is often used to measure this independence and determines the degree to which the rate and depth of fading may be reduced. The cross-correlation coefficient varies for different antenna spacing and different directions and beamwidths of incoming signals. It is generally accepted that reasonable improvement in received signal statistics can be achieved with a cross-correlation coefficient of seven-tenths. The required spacing at the base station differs for spatial separation in the horizontal and vertical planes. For horizontal spacing, the separation has been found to depend not only on the angle between the line joining the base station and mobile and the line joining the base station antennas, but also on the above-ground height of the antennas. In addition, the presence of local scatters in the immediate vicinity of the base station antennas will have an impact. It has been quoted that the horizontal space diversity at a base station requires antenna spacing of up to about 20λ [11.4]. Vertical space diversity requires a separation of about

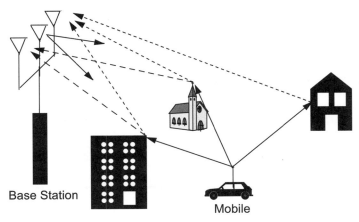

Figure 11-6 Space Diversity in Multipath Environment

15λ; this also depends on the effective scattering radius of the area in which the mobile is located. Besides, an increase in the base station antenna height results in a cross-correlation decrease, providing that the base station antenna spacing remains unchanged.

The space diversity gain is about 3–5 dB for the horizontal separation and 2–4 dB for the vertical separation. Horizontal space diversity gives a better performance than that of the vertical separation because the de-correlation of the received signals increases faster with the horizontal rather than with the vertical separation of the antennas.

When the transmit antenna (mobile or handset antenna) is tilted from the vertical direction, the received signal level reduces on both antenna branches because the vertically polarized component of the transmitted signal reduces as the tilt angle increases. Although there is a loss in the signal level with a tilt, the diversity gain in a space diversity does not change, which means that the fading depth will still be reduced with diversity as the transmit antenna is tilted.

11.4.2 Polarization Diversity Antenna

Orthogonally polarized antennas at a base station provide polarization diversity, a means of realizing two independently fading signals without the need for physically separate antennas as in space diversity (see References [11.4-6]). In PCS/cellular systems, polarization diversity is likely to become more important because it employs small cells and antenna heights comparable to, or lower than, the surroundings, where it may be difficult to mount two antennas with the appropriate spacing for space diversity.

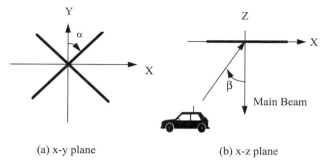

(a) x-y plane (b) x-z plane

Figure 11-7 Polarization Diversity Antenna at the Base Station

Figure 11–7 shows the polarization diversity antenna system coordinates. As shown in Figure 11–7(a), when the polarization angle, α, is zero, the antenna system provides V-H polarization diversity. However, when $\alpha = 45$ degrees, a slant forty-five-degree polarization diversity can be obtained. Figure 11–7(b) shows a mobile user's location in the direction of offset angle β from the main beam direction of the antennas.

For polarization diversity antenna, as shown in Figure 11–7, the cross-correlation coefficient ρ can be expressed as

$$\rho(\alpha,\beta,\Gamma) = \frac{\tan^2(\alpha)[\Gamma - \cos^2(\beta)]^2}{[\tan^2(\alpha)\cos^2(\beta) + \Gamma][\Gamma\tan^2(\alpha) + \cos^2(\beta)]}, \quad (11.9)$$

and the average value of signal loss, L, relative to that received using vertical polarization is given by

$$L(\alpha,\beta,\Gamma) = \frac{\Gamma + \tan^2(\alpha)\cos^2(\beta)}{\Gamma\tan^2(\alpha) + \cos^2(\beta)}, \quad (11.10)$$

where $\Gamma = \langle r_2 \rangle / \langle r_1 \rangle$ is the cross-polarization discrimination of the propagation path between a user and a base station and r_1 and r_2 are two independent random variables with Rayleigh distribution. From (11.9) and (11.10), it can be seen that both cross-correlation coefficient and signal loss are determined by three factors: polarization angle α, offset angle β, and cross-polarization discrimination Γ.

Figure 11–8 shows the three dimensional plots of the calculated cross-correlation coefficient ρ and received signal level decrease L as a function of α and Γ. The range of α is from zero to forty-five degrees. In Figures 11–8(a) and (b), ρ generally becomes higher as both polarization angle α and cross-polarization discrimination Γ become larger. ρ increases when offset angle β increases, and there is no diversity gain when $\beta = 90$ degrees ($\rho = 1.0$ from Equation (11.9)). This is because the received signal on

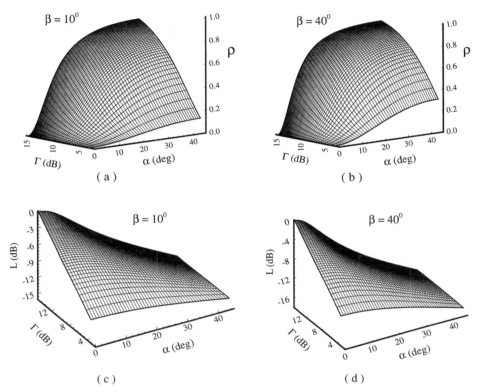

Figure 11–8 Cross-Correlation Coefficient and Received Signal Level Decrease as a Function of Polarization Angle and Cross-Polarization Discrimination for Two Kinds of Offset Angle

two branches is only a vertical polarization component at $\beta = 90$ degrees. In Figures 11–8(c) and (d), L becomes larger as α increases or as Γ decreases. Additionally, L increases as β increases because the horizontally polarized component becomes smaller as β increases.

Measurements have shown that a cross-correlation coefficient, with H-V and a forty-five-degree slant polarization diversity, is lower than seven-tenths in urban, suburban, and rural areas at PCS band [11.7]. The cross-polarization discrimination is around 10 dB in urban and suburban areas and 11 to 12 dB in other areas.

Table 11-2 Correlation Coefficients at Ninety Percent Signal Reliability for the Various Testing Configurations

Type of Transmitting Antenna	Space Diversity	V-H Pol. Diversity	Slant 45° Pol. Diversity
Vertical Mobile	0.60	0.25	0.68
Slant Mobile	0.54	0.53	0.43
In Car Random Orientation	0.57	0.44	—

11.4.3 Comparison of Polarization Diversity and Space Diversity

Comparison of space diversity and polarization diversity shows the following observations.

- There can be a power loss (for example, 3 dB) in the forward-link signal of the polarization diversity since the transmitter power is split into the two polarizations.
- When the transmitter antenna is in vertical direction, the cross-correlation coefficients obtained for polarization diversity in any environment are less than seven-tenths at ninety percent signal reliability, which is comparable to those obtained for horizontal separation.
- In the case of a slanted mobile transmitting antenna, a forty-five-degree slant polarization diversity antenna at the base station shows better correlation statistics than a space diversity antenna. Table 11–2 shows the comparisons of the correlation coefficients for three kinds of mobile antennas and base station antennas.
- The diversity gain of the polarization diversity antenna is about 2 dB when the mobile antenna is vertical. However, tilting the mobile antenna results in a gain increase from 3 to 5 dB.
- The specific performance of the polarized diversity antenna relative to space diversity is a function of the amount of reflection and/or scattering in the local environment. Field measurements suggest that the use of polarized diversity antennas in certain morphologies may result in performance loss of up to 2 dB.

11.5 Antenna Isolation Guidelines for Collocated RF Stations

11.5.1 Introduction

Because of deployment constraints, estate acquisition difficulties, and other reasons, it is sometimes highly desirable that the CDMA PCS cell site be collocated with RF stations of other communications systems such as TDMA PCS, cellular CDMA, cellular TDMA, AMPS, AM, SMR, etc. When they are collocated, mutual interference between stations always exists, which may cause receiver desensitization, overload, and/or intermodulation product (IMP) interference, thereby degrading their system performances. Therefore, if the service provider wants to collocate a CDMA PCS cell site with other RF station(s), precautions should be taken to avoid/minimize harmful mutual interferences.

The degree of degradation depends on the strength of the interfering signal, which transmit/receive (TX/RX) unit performance, spectrum spacing, and antenna separation between the collocated stations determine. This section will first introduce a set of mathematical models necessary for evaluating the mutual interference between two collocated RF communications stations and will specify the criteria for determining the required antenna isolations between the collocated stations. The models will then be generalized for the cases in which more than two RF stations are collocated (called *multiple collocated RF stations* hereafter). Finally, site survey, which is an important step in the process of collocating a CDMA PCS cell site with any other RF station(s), will then be explained.

11.5.2 Mathematical Model for Mutual Interference Evaluation

Figure 11–9 is a schematic diagram showing the mutual interference between two collocated RF stations. The RF components that are very important in evaluating the mutual interference between two collocated stations are the TX amplifier and TX filter of the interfering station and the RX filter and receiver (that is, preamplifier) of the interfered station as indicated in Figure 11–9.

Note that the term *antenna isolation* (called *isolation* hereafter) refers to the path loss between the antenna terminals (that is, input/output port of the RX/TX unit) of the collocated stations. This includes the propagation loss through the air and the effective antenna gains (for example, antenna gain plus cable loss) of both stations.

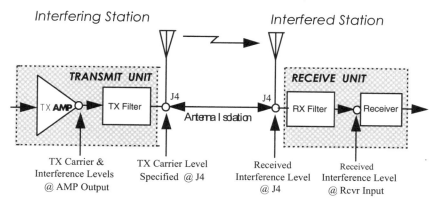

Figure 11-9 Mutual Interference Between Two Collocated RF Stations

Three kinds of degradation need to be considered:

1. receiver desensitization,
2. IMP interference, and
3. receiver overload.

A strong spurious emission received from the interfering stations causes receiver desensitization. A combination of carriers received from the collocated stations generates the IMP interference. The fact that total power received is too strong accounts for receiver overload. To minimize those degradations, without modifying the existing TX/RX units, appropriate *isolations* need to be maintained between the collocated stations. Use of the criteria set forth in Section 11.5.3 will determine the required isolations.

11.5.2.1 Receiver Desensitization

Receiver desensitization is the degradation in receiver sensitivity due to an increase in the *receiver noise floor*. In this discussion, we assume that the TX band of the interfering station is *adjacent* to the RX band of the interfered station. Therefore, the interfering station generates a substantial amount of spurious emission. If the isolation between two stations is insufficient or if the interfering station's TX filter does not provide enough out-of-band attenuation (that is, rejection), the spurious emissions that fall into the RX band of the interfered station may be too strong, resulting in an increase in the receiver noise floor.

It can be imagined from Figure 11-9 that the interference power due to spurious emissions generated at the TX amplifier output of the interfering station is filtered by the TX filter, attenuated by the isolation between two stations, and then received by the RX unit at the interfered station. Therefore, the affected interference power received at the antenna terminal of the interfered station can be expressed as

$$I_{aff\_J4} = C_{TX\_amp} + ICR_{TX\_amp} - L_{TX\_rej} - \mathcal{L}_{isolation}^{spurious} + 10\log_{10}(W_{interfered}/W_{interfering}), \quad (11.11)$$

where

I_{aff_J4} = affected interference level at interfered station's antenna terminal (dBm),

C_{TX_amp} = nominal maximum *carrier* power level at TX amplifier output (dBm),

ICR_{TX_amp} = interference-to-carrier ratio of TX amplifier (dBc),

L_{TX_rej} = TX filter rejection of *signal* in interfered station's RX band (dB),

$\mathcal{L}_{isolation}^{spurious}$ = isolation between antenna terminals of collocated stations (dB),

$W_{interfered}$ = interfered system channel bandwidth (KHz), and

$W_{interfering}$ = interference level measurement bandwidth (KHz).

Swapping parameters I_{aff_J4} and $\mathcal{L}_{isolation}^{spurious}$, we can express (11.11) in an alternate form as

$$\mathcal{L}_{isolation}^{spurious} = C_{TX\_amp} + ICR_{TX\ amp} - L_{TX\_rej} - I_{aff\_J4} + BWAF. \quad (11.12)$$

Note that the BWAF is the *Bandwidth Adjustment Factor*, which is defined as

$$BWAF = 10\log_{10}(W_{interfered}/W_{interfering}). \quad (11.13)$$

If the level of TX carrier power is specified at the antenna terminal (denoted as C_{TX_J4}, in dBm), the C_{TX_amp} can be calculated by

$$C_{TX\_amp} = C_{TX\_J4} + L_{TX\_pass}, \quad (11.14)$$

where L_{TX_pass} is the TX filter passband loss. Substituting (11.14) into (11.12), we have

$$\mathcal{L}_{isolation}^{spurious} = C_{TX\_J4} + ICR_{TX\_amp} + (L_{TX\_pass} - L_{TX\_rej}) - I_{aff\_J4} + BWAF. \quad (11.15)$$

11.5.2.2 *Intermodulation Product Interference*

Due to the non-linearity of the receiver gain transfer function, if the strengths of received interfering carriers are too strong, the IMPs will be generated in the receiver (that is, preamplifier) and presented at its output. Sometimes these IMPs may be too strong. If they fall into the RX band of the interfered system, they may cause interference and degrade receiver performance.

Antenna Isolation Guidelines for Collocated RF Stations

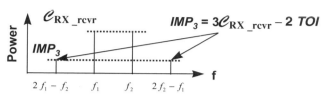

Figure 11-10 Third-Order IMP Generated by Two Input Signals in the RX Path

Many studies have been conducted on the nature of IMPs and show that third-order IMPs (IMP3) are the stronger ones and that they may have adverse effects on receiver performance. Therefore, in the following discussions, we will focus on the IMP3.

Mathematically, the IMP3 (in dBm), which is generated by two carriers of equal strength (called *two-tone IMP3*), can be calculated by

$$\text{IMP3} = 3 \times \mathcal{C}_{\text{RX\_rcvr}} - 2 \times \text{TOI}, \tag{11.16}$$

where TOI = Third-Order Intercept point specified at receiver input (dBm) and $\mathcal{C}_{\text{RX\_rcvr}}$ = interference carrier level at interfered station's receiver input (dBm). Figure 11-10 illustrates the relationship of (11.16).

Note that to minimize receiver degradation caused by IMP3 interference, the strengths of IMP3s shall be lower than an acceptable level, which will be specified in Section 11.5.3. From (11.16), the $\mathcal{C}_{\text{RX\_rcvr}}$ for achieving a specified IMP3 level can be obtained by

$$\mathcal{C}_{\text{RX\_rcvr}} = \frac{\text{IMP3} + 2 \times \text{TOI}}{3}. \tag{11.17}$$

Referring to Figure 11-9, we can also express the $\mathcal{C}_{\text{RX\_rcvr}}$ as

$$\mathcal{C}_{\text{RX\_rcvr}} = C_{\text{TX\_J4}} - \mathcal{L}_{\text{isolation}}^{\text{IMP3}} - L_{\text{RX\_cxr\_rej}}, \tag{11.18}$$

where

$C_{\text{TX\_J4}}$ = nominal maximum TX carrier power at interfering station's antenna terminal (dBm),

$\mathcal{L}_{\text{isolation}}^{\text{IMP3}}$ = isolation between antenna terminals for achieving a specified $\mathcal{C}_{\text{RX\_rcvr}}$ (dB), and

$L_{\text{RX\_cxr\_rej}}$ = RX filter rejection of carrier received from interfering station (dB).

Swapping $\mathcal{L}_{isolation}^{IMP3}$ and $\mathcal{C}_{RX\_rcvr}$, we have

$$\mathcal{L}_{isolation}^{IMP3} = C_{TX\_J4} - L_{RX\_cxr\_rej} - \mathcal{C}_{RX\_rcvr}. \tag{11.19}$$

Note that the specified IMP3 level determines the $\mathcal{C}_{RX\_rcvr}$ as indicated in (11.17).

11.5.2.3 Receiver Overload

Excessively strong received power at the receiver causes receiver overload. When a receiver is driven into overload, its amplification gain is decreased (that is, depressed). To prevent the receiver from being overloaded, the level of *total* carrier power received from the interfering station needs to be well below its 1 dB compression point (P_{1dB}). Section 11.5.3 will specify the acceptable level.

Referring to Figure 11-9 and following the analysis similar to the one in the previous section, we can express the isolation (denoted as $\mathcal{L}_{isolation}^{overload}$) between two collocated stations to suppress the level of total affected carrier power received at the interfered station's receiver to an acceptable level as

$$\mathcal{L}_{isolation}^{overload} = C_{TX\_tot\_J4} - L_{RX\_cxr\_rej} - \mathcal{C}_{RX\_tot\_rcvr}, \tag{11.20}$$

where

$C_{TX\_tot\_J4}$ = total carrier power transmitted at antenna terminal of interfering station (dBm),

$\mathcal{C}_{RX\_tot\_rcvr}$ = total carrier power received at antenna terminal of interfered station (dBm),

$\mathcal{L}_{isolation}^{overload}$ = isolation between antenna terminals for achieving a specified $C_{RX\_tot\_rcvr}$ (dB), and

$L_{RX\_cxr\_rej}$ = RX filter rejection of carrier power received from interfering station (dB).

Note that every receiver (that is, preamplifier) is designed to operate properly within a specific bandwidth called the *operation band*. If the received signal falls into the operation band, its strength will be amplified; otherwise, it will be attenuated. Therefore, the receiver behaves as an active band-pass filter in a sense because it has uniform gain (for example, 18 dB) within its operation band and high attenuation (that is, loss) outside the band. The degree of attenuation depends on the receiver design and the frequency difference between the carrier and the operation band. In some cases, the receiver may become inactive to the incoming carriers whose frequencies are only a few tens of MHz outside its operation band; therefore, it seems that the receiver does not

Antenna Isolation Guidelines for Collocated RF Stations   193

receive these carriers. In other words, the receiver *totally rejects* the power of these incoming carriers. Due to this fact, we point out that (11.20) applies only under the condition that the collocated stations are operating in the adjacent bands (for example, PCS at A-Block and PCS at D-Block) in which the receiver of the interfered station is active to the carriers from the interfering stations.

11.5.3 Antenna Isolation Criteria and Safe Antenna Isolation

To ensure proper system performance, the three types of degradation mentioned above should be either avoided or minimized. To achieve this goal, the antenna isolation between the collocated stations will be maintained to meet the criteria given below.

1. *At the interfered station, the strength of affected spurious emissions received from the interfering station(s) will be 10 dB below its receiver noise floor.*
2. *At the interfered station, the level of IMP3 products generated will be 10 dB below its receiver noise floor.*
3. *At the interfered station, the strength of the total affected carrier power received from the interfering station(s) will be 5 dB below its receiver 1 dB compression point (P_{1dB}).*

Note that to meet all the criteria listed above, the largest antenna isolation will be chosen. In most cases, criterion 1 determines the largest antenna isolation. If we use this antenna isolation, the RX sensitivity of the interfered system is degraded only by 0.5 dB, which is considered acceptable for most communications systems.

As mentioned at the beginning of this section, mutual interference always exists between the collocated stations. For example, when considering two collocated RF stations, say Station A and Station B, the TX unit of Station A will affect the RX unit of Station B, and vice versa. Therefore, two sets of required antenna isolations, one for the interference from Station A to Station B and the other one for the interference from Station B to Station A, need to be calculated. Similarly, to avoid/minimize the degradation to both stations, the largest antenna isolations (called *safe antenna isolation* (SAI)) obtained in these two sets shall be utilized. Using the SAI, the maximum receiver degradation of both stations is approximately 0.5 dB.

Example: Antenna Isolation

In this example, we show how application of the interference models and criteria explained above finds appropriate antenna isolation between two collocated systems. Suppose that we want to know the required isolation between a TDMA system in the A Block and a CDMA system in the B Block. We will assume that the impact of the interference generated by the TDMA system on the CDMA system is more serious than that generated by the CDMA system on the TDMA system and consider only the interference generated by the TDMA system. Also, we assume that

- the noise floor of the affected CDMA system is -108 dBm in 1.25 MHz,
- the receiver 1 dB compression point and TOI of the CDMA system are, respectively, -18 dBm (P_{1dB}) and -6 dBm,
- CDMA RX filter attenuation in the TDMA TX band is 84.8 dB ($L_{RX\_cxr\_rej}$),
- per-carrier TDMA amplifier output power is 42.8 dBm (C_{TX_amp}) and the interference-to-carrier ratio at the amplifier output is -80 dBc/30 KHz ($ICR_{TX\ amp}$),
- TDMA TX filter attenuation in the CDMA RX Band is 51.5 dB (L_{TX_rej}),
- TDMA carrier power at the antenna terminal is 42 dBm (C_{TX_J4}), and
- TDMA total carrier power at antenna terminal is 60.6 dBm/15 MHz ($C_{TX\_tot\_J4}$); frequency reuse of seven is used so that the total number of 30 KHz TDMA carriers per base station is about seventy-two (=(15MHz/30KHz)/7).

(i) <u>Receiver Desensitization</u>:

The acceptable interference, which is 10 dB below the noise floor, becomes -118d Bm/1.25 MHz. Since BWAF = $10 \log(1.25 \times 10^6 / 30 \times 10^3)$ = 16.2 dB from (11.12), the required isolation for spurious emission becomes

$$\mathcal{L}_{isolation}^{spurious} = 42.8 + (-80) - 51.5 - (-118) + 16.2 = 45.5 \text{ dB}.$$

(ii) <u>Third-Order Intermodulation Product (IMP)</u>:

The acceptable IMP3 power at the CDMA base is 10 dB below the noise floor, that is, -118 dBm/1.25 MHz. Therefore, from (11.17), the acceptable TDMA carrier power received at the CDMA system is

$$\mathcal{C}_{RX\_rcvr} = [(-118) + 2(-6)] / 3 = -43.3 \text{dBm}.$$

Then from (11.19),

$$\mathcal{L}_{isolation}^{IMP3} = 42 - 84.8 - (-43.3) = 0.5 \text{dB}.$$

(iii) Receiver Overload:

The acceptable total carrier power at the CDMA base is -23 dBm ($\mathcal{C}_{RX\_tot\_rcvr}$), which is 5 dB below -18 dBm, the receiver 1 dB compression point. Therefore, from (11.20),

$$\mathcal{L}_{isolation}^{overload} = 60.6 - 84.8 - (-23) = -1.2 \text{ dB}.$$

We find, in this example, that criterion 1 (used to ensure the system received spurious emissions 10 dB below the receiver noise floor) is the dominant factor in determining the antenna isolation guideline. ❑

We provide isolation estimates as starting points. Since it is difficult to predict actual path loss by the propagation model, measurement data should be taken to ensure achievement of antenna isolation. Calculated isolation estimates may change with further refinements in the method of analysis.

11.5.4 Antenna Separation between Two Collocated RF Stations

Antenna isolation, $\mathcal{L}_{isolation}$ (in dB), between the antenna terminals of two collocated stations can be expressed as

$$\mathcal{L}_{isolation} = \mathcal{L}_{prop}(d) - G_{interfering}^{effect} - G_{interfered}^{effect}, \quad (11.21)$$

with

$$G_{interfering}^{effect} = G_{interfering}^{TX\_antenna} - L_{interfered}^{TX\_cable}$$

$$G_{interfered}^{effect} = G_{interfered}^{RX\_antenna} - L_{interfered}^{RX\_cable},$$

where

$\mathcal{L}_{prop}$ = propagation loss between collocated stations (dB),

$G_{interfering}^{TX\_antenna}$ = interfering station TX antenna gain *in interfered station RX band* and *in the direction pointing to the interfered station* (dBi),

$G_{interfered}^{RX\_antenna}$ = interfered station RX antenna gain *in its RX band* and *in the direction pointing to the interfering station* (dBi),

$L_{interfering}^{TX\_cable}$ = the interfering station's TX path cable loss (dB), and

$L_{interfered}^{RX\_cable}$ = the interfered station's RX path cable loss (dB).

Note that the $G_{\text{interfering}}^{\text{TX\_antenna}}$ and $G_{\text{interfered}}^{\text{RX\_antenna}}$ are the actual antenna TX/RX gains that take into account all the reflection, diffraction, and/or scattering caused by the support structures that are close to the stations. Therefore, the $G_{\text{interfering}}^{\text{TX\_antenna}}$ and $G_{\text{interfered}}^{\text{RX\_antenna}}$ may not be the same as those specified by the antenna manufacturer. The antenna gain is conventionally specified in dBi, which is a relative quantity used to express the difference in gain between the antenna under test and an isotropic radiator.

If there is line-of-sight (LOS) between the collocated stations and the free-space propagation loss model is applied, Equation (11.21) can be expressed explicitly as

$$\measuredangle_{\text{isolation}} = [20 \log_{10}(f) + 20 \log_{10}(d) - 37.93] - G_{\text{interfering}}^{\text{effect}} - G_{\text{interfered}}^{\text{effect}} \quad (11.22)$$

where f = frequency of affecting interference (spurious emission or carrier) (MHz) and d = antenna separation between two collocated stations (feet). Therefore, if f, $G_{\text{interfering}}^{\text{effect}}$, and $G_{\text{interfered}}^{\text{effect}}$ are known, the antenna separation d to achieve the required isolation $\measuredangle_{\text{isolation}}$ can be determined by the following equation:

$$d = 10^{\text{SUM}/20}, \quad (11.23)$$

with SUM = $\measuredangle_{\text{isolation}} + G_{\text{interfering}}^{\text{effect}} + G_{\text{interfered}}^{\text{effect}} + 37.93 - 20 \log_{10}(f)$.

Note that (11.23) is useful only under the condition that the free-space propagation loss model is valid (for example, LOS) and $G_{\text{interfering}}^{\text{effect}}$ and $G_{\text{interfered}}^{\text{effect}}$ can be accurately estimated. However, in reality, due to the effects caused by nearby antenna support structures mentioned above, it is very difficult to generate a mathematical model to accurately predict the antenna pattern and its gain. Therefore, we recommend that the required antenna separation be experimentally determined.

Example: Antenna Separation

In the example illustrated in the previous section, the required antenna isolation was found to be 45.5 dB. Assuming that TDMA TX antenna gain minus cable loss is 10 dBi ($G_{\text{interfering}}^{\text{effect}}$), CDMA RX antenna gain minus cable loss is 10 dBi ($G_{\text{interfered}}^{\text{effect}}$), and the CDMA RX frequency is 1877.5 MHz (B Block), we find from (11.23) that the isolation corresponds to the antenna separation given by

$$d = 10^{[(45.5 + 10 + 10 + 37.93 - 20\log(1877.5)]/20} = 79 \text{ feet} \qquad \square$$

11.5.5 Mutual Interference Between Multiple Collocated RF Stations

Having sufficient knowledge about how to deal with two collocated stations, now we are ready to expand the discussion to include the interference among more than two

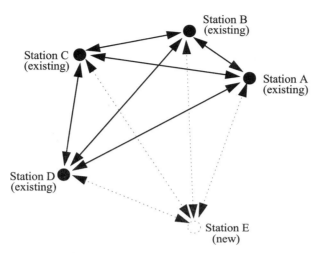

Figure 11–11 Mutual Interference Between Five Collocated RF Stations

collocated RF stations. For illustration purposes, we only consider five collocated RF stations, as shown in Figure 11–11. However, due to a generalized description/explanation, the concepts and conclusions provided herein can be extended and applied to the cases having any number of collocated stations.

As indicated in Figure 11–11, there are four existing stations, denoted as Stations A, B, C, and D, in the site area, and a service provider is planning to install a new station (that is, Station E) at the location specified by the dashed circle. The solid lines represent the mutual interference between the existing stations, whereas the dashed lines indicate the mutual interference between the new station and the existing stations. To simplify the discussion, we assume that, before installing the new station, all the existing stations were functioning normally and the interference received at each existing station was at its acceptable level.

Two factors mainly determine whether to allow the new station to collocate with the existing stations: (i) inclusion of the new station will not degrade the RF performance of all the existing stations and (ii) the levels of total interference received from *all* the existing stations will meet the three criteria for the new station, as set forth in Section 11.5.3.

Due to difficulties in gathering the required TX/RX unit specifications for all of the existing stations, it is a cumbersome and time-consuming task to deal with the multiple stations collocation issue. The degree of difficulty depends on the number of collocated stations considered. In this section, based on Figure 11–11, we will briefly describe the approaches for determining a location for the new station (that is, Station E) in the collocation area.

11.5.5.1 *Interference from a New Station to Existing Stations*

As explained at the beginning of 11.5.5, to evaluate the interference from the new station (that is, Station E) to each existing station, we need the following information: the TX unit performance specifications of the new station and the RX unit performance specifications of all existing stations.

The installation of the new station in the collocated area is under the condition that all the existing stations have sufficient sensitivity margins to tolerate/accept additional interference coming from the new station without causing any undesired degradation. To limit the additional interference to an acceptable level, we will maintain appropriate antenna isolation between the new station and each existing station.

11.5.5.2 *Interference from Existing Stations to New Station*

To determine the interference from all the existing stations to the new station, we need the following information: the RX unit performance specifications of the new station and the TX unit performance specifications of all existing stations.

To evaluate the degradation at the new station, the total interference coming from all the existing stations needs to be considered. To meet the criteria set forth in Section 11.5.3, another set of required antenna isolations between the new station and each existing station needs to be maintained. Note that those antenna isolations may be different from those obtained in 11.5.5.1. To ensure the interference received at the new station and all the existing stations is at their acceptable levels, the safe antenna isolation, which is the largest of antenna isolations obtained in 11.5.5.1 and 11.5.5.2, between the new station and each existing station must be chosen.

11.5.5.3 *Antenna Isolation/Separation Estimations*

Due to difficulties in obtaining all the TX/RX unit performance specifications and accurately predicting all the antenna gains/patterns in the field, it is impossible to derive a mathematical equation to estimate the required antenna isolations/separations between the new and all of the existing stations. Due to lack of sufficient information and to ensure proper operation for all the collocated stations under study, we usually add and/or make a number of assumptions to include relatively large margins on the TX/RX unit performance (for example, TX/RX filter rejection factor, etc.). Those margins may be too conservative, thereby resulting in over-estimated antenna isolation results, which are much larger than those actually required. For this reason, we recommend that, to achieve more accurate and reasonable results, the antenna isolations/separations be determined experimentally.

11.5.6 Site Survey

As explained above, the accurate and meaningful results can only be achieved through direct measurement. Site survey (SS) is an approach to fulfill this measurement. It is a very important step in the process of setting up a new station (for example, CDMA PCS cell site) and therefore must be performed after the location is selected and before the equipment is installed for the new station, whether it is *stand-alone* or *collocated with other stations*. We will briefly describe the tasks to be performed in the SS. In the following discussion, we assume that the new station is a CDMA PCS cell site.

The major tasks that should be completed in the SS are listed below.

(1) Measure the *frequencies and strengths* of all the existing carriers and perform an IMP3 study to evaluate the IMP3 interference before and after the new station is installed.
(2) Measure the strengths of *spurious emissions* falling in the new station's RX band.
(3) Perform a receiver overload study based on the frequencies and strengths of carriers measured in (1).
(4) Measure electrical field (E-field) intensity at the equipment frame (optional for high-power interference case).

Note that all the measurements given above must be performed at the location selected for the new cell site. To achieve more accurate results, we suggest that the same type of RX/TX antenna chosen for the CDMA PCS cell site be utilized for measurements (1) and (2) listed above. To closely simulate the real situation, the antenna must be lifted as high as possible to the height proposed for it and its main beam must be pointing in the desired direction.

We suggest using the exact RX/TX antenna chosen for the new CDMA PCS cell site in the measurements because the antenna is a *frequency sensitive device*; its gain, radiation pattern, and terminal impedance (called *antenna impedance*) change with frequency. The antenna manufacturer provides the performance of these quantities (called *antenna specifications*).

These specifications are valid only within the bandwidth (that is, *operation band*) designed for the antenna. Normally it is very narrow, about five to ten percent of its central frequency. Within the operation band, the antenna performance meets the nominal specifications. However, when operating outside the band, the antenna performance deteriorates very rapidly; the antenna's gain and radiation pattern are degraded in an unpredictable way, and its terminal impedance becomes mismatched to the feeding system. If a mismatch becomes very high, which usually occurs at frequencies much higher or lower than the band of operation, the signal is reflected and no power can be transmitted and received by the antenna. Since out-of-band performance information is not important for the normal use of an antenna, it is usually not included in the specifications provided for the customers.

Based on the reasons outlined above, it is obvious that when the collocated stations are operating in different bands, the frequencies of which are far apart (for example, PCS at 1900 MHz, cellular at 850 MHz, AM at 1 MHz, etc.), it is incorrect and meaningless to use the *in-band* antenna gain/pattern information (that is, specifications) provided by the antenna manufacturer to evaluate the *out-of-band* interference effect. The antenna specifications may be useful only when dealing with the collocated stations that are operating in the same or adjacent bands.

The tasks of SS outlined at the beginning of this section are the easy, correct, and straightforward approaches for evaluation of interference received from the *known* and/or *unknown* stations. The *known stations* are the stations whose locations are explicitly known, whereas the *unknown stations* are the hidden stations whose locations are not known. Note that sometimes the interference and jamming coming from the hidden station(s) may be the strongest and therefore require more attention.

Task 1—IMP3 Interference

The purpose of Task 1 is to evaluate the difference in the IMP3 interference in the collocated site *before* and *after* the new station is installed. To determine the IMP3 interference, we shall first measure the frequencies of all the existing carriers at the location chosen for the new station. Using the frequency data obtained in the measurement, the frequencies of all the IMP3s are calculated and the IMP3 interference can then be evaluated.

As mentioned earlier, the IMPs are generated due to non-linearity of the gain transfer function. The general form of the non-linear transfer function for a receiver can be expressed by an infinite power series as

$$e_{out} = \sum_{n=1}^{\infty} a_n e_{in}^n, \tag{11.24}$$

where a_n are coefficients and e_{in} and e_{out} are the carriers at input and output of the receiver, respectively.

Antenna Isolation Guidelines for Collocated RF Stations

Here we assume that the transfer function is *quasi-linear,* which implies that the coefficient a_1 in (11.24) is much larger than the coefficients a_2 through a_∞. Furthermore, if e_{in} is sufficiently small, the infinite series in (11.24) may be approximated by the first three terms as

$$e_{out} = a_1 e_{in} + a_2 e_{in}^2 + a_3 e_{in}^3 . \qquad (11.25)$$

The IMP3s are generated by the third term in (11.25). For any three carrier frequencies (for example, α, β and γ), a set of IMP3s will be generated and their frequencies are given below:

$$(3\alpha), (3\beta), (3\gamma)$$

$$(2\alpha + \beta), (2\alpha - \beta), (2\alpha + \gamma), (2\alpha - \gamma),$$
$$(2\beta + \alpha), (2\beta - \alpha), (2\beta + \gamma), (2\beta - \gamma),$$
$$(2\gamma + \alpha), (2\gamma - \alpha), (2\gamma + \beta), (2\gamma - \beta),$$

$$(\alpha + \beta + \gamma), (\alpha + \beta - \gamma), (\alpha - \beta + \gamma), (\alpha - \beta - \gamma).$$

Since all the receivers have a transfer function similar to that given in (11.25) and since every three carriers will generate a set of IMP3 frequencies as listed above, all the IMP3 frequencies existing at the collocated site can be determined. This is done by using the measured carrier frequencies obtained in Task 1 of the site survey before installation of the new station. Since having been harmonically collocated, it is a reasonable assumption that these IMP3 frequencies will not fall in any of existing stations' RX band.

If the new station is installed, a combination of existing carrier(s) and new carrier(s) from the new station will generate additional IMP3s and exponentially increase their number. Note that the more IMP3s they generate, the greater the chance that the collocated stations will be polluted by them. In an ideal situation, the new IMP3 frequencies do not fall into the RX band of any collocated station, including the existing stations and new station. When this becomes true, the work of Task 1 is completed.

However, if some of IMP3 frequencies fall into the new station's RX band, the causes (that is, affecting carriers) of these damaging IMP3 products need to be identified and their strengths determined. Since it is very difficult to generate a mathematical model to accurately estimate the strength of IMP3s, we recommend that, if this situation occurs, the IMP3 strength be determined through direct measurement.

Simply attaching the actual RX filter and receiver at the other end of the antenna cable (that is, the antenna terminal) performs the measurement. If the affecting carriers exist outside the RX band, due to the extra rejection provided by the RX filter, their strengths may be suppressed dramatically, thereby substantially reducing the levels of damaging IMP3s. Note that in general a 1 dB reduction in the carrier levels results in a 3 dB drop in IMP3 power. Therefore, all the affecting IMP3s will be suppressed to the acceptable level with the use of the RX filter.

Task 2—Spurious Emissions

The purpose of Task 2 is to measure the strength of the total spurious emissions received from all the existing stations. Simply attaching a spectrum analyzer at the antenna terminal, setting an appropriate resolution bandwidth (for example, 1 KHz), and tuning the scanning bandwidth to cover the entire new station's RX band performs this measurement. Note that the level of *total spurious emissions* measured must meet criterion 1 set forth in Section 11.5.3.

Task 3—Receiver Overload

The *total* power of all the carriers received at the receiver's input determines the receiver overload. Attaching the exact RX filter between the antenna terminal and spectrum analyzer accurately determines each carrier's power strength. Note that the total power, which is the summation of all the carrier power measured, must meet criterion 3 set forth in Section 11.5.3.

Task 4—Electrical Field Intensity Measurement

The purpose of this task is to determine the intensity of the electrical field (E-field, in volts/meter) generated by the strong interfering carrier. The E-field intensity is an important quantity in evaluating the capability or possibility of installing the equipment frame close to a station that is transmitting a very high power (for example, a thousand watts). Therefore, when planning to locate the CDMA PCS cell site close to an active AM station/tower, this measurement is necessary.

Table 11–3 Field Intensity Specifications

Standard	E-Field Intensity (V/m)	Frequency Range
IEC 1000-4-3	3	27 MHz - 1000 MHz
Bellcore GR-1089-CORE	10	10 KHz - 10 GHz

To achieve a large coverage (for example, R = 40 miles), the AM station normally transmits a very high power (for example, a thousand watts or more). The AM station usually consists of an array of monopoles to create a specific radiation pattern. Those monopoles use the earth's surface as their ground plane. Most of electromagnetic waves (that is, energy/power) launched by these monopoles are supported by the earth's surface (called *ground waves*) and propagate away horizontally. Therefore, the strongest field intensity is concentrated at a level/height very close to the earth's surface where the CDMA PCS cell site equipment frames (cabinets) are normally installed. To prevent equipment from being damaged by a high-intensity field, the E-field measured at the equipment frame's surface must be below the limits specified in Table 11–3. If the E-field intensity is stronger than that specified in Table 11–3, we must either relocate the new station or cover the equipment frame with extra shielding materials.

11.6 References

[11.1] C. A. Balanis. *Antenna Theory Analysis and Design*, Harper & Row, Publication Inc., 1982.

[11.2] W. L. Stuzman and G. A. Thiele. *Antenna Theory and Design*, John Wiley & Sons, Inc., 1981.

[11.3] E. C. Jordon. *Reference Data for Engineerings: Radio, Electronics, Computer, and Communications*, H. W. Sams & Co., Macmillan Inc., 1992.

[11.4] W. C. Y. Lee. *Mobile Communications Engineering*, McGraw-Hill Company, 1982.

[11.5] R. G. Vauhan. Polarization Diversity in Mobile Communications, *IEEE Trans. Vehicular Technology*, vol. 39, no. 3, pp. 177-186, 1990.

[11.6] W. Honcharenko. Measurements Comparing Space Diversity, VH Polarization Diversity, and 45° Slant Polarization Diversity, *Technical Memorandum*, Lucent Technologies, 1996.

[11.7] A. M. D. Turkmani, A. A. Arowojolu, P.A. Jefford, and C. J. Kellett. An Experimental Evaluation of the Performance of Two-Branch Space and Polarization Diversity Schemes at 1800 MHz, *IEEE Trans. Vehicular Technology*, vol. 44, no. 2, pp. 318-326, 1995.

APPENDIX A

RF Design Process

A.1 Purpose

This appendix describes the basic or fundamental practices, techniques, and degrees of freedom available to a RF design engineer designing the RF portion of a new PCS network to meet the requirements of service quality, coverage, and capacity as specified by the PCS service provider. This appendix describes a generic RF design procedure.

A.2 Process Overview

The PCS system RF design process consists of two phases: preliminary design and final design (see Figure A–1).

The preliminary design phase the RF design of a new network. The major aspects of this phase are:

- gathering and synthesizing inputs to characterize the regions in which the new network must operate,
- using limited drive-test results to tune the prediction tool to make a rough prediction of the quantity of cell sites potentially needed to serve a targeted region, and
- directing the site acquisition process to identify candidates within the search area rings.

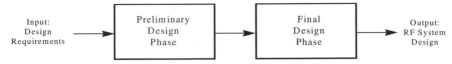

Figure A–1   PCS System RF Design Process

The purpose of the final design phase is to refine and finalize the RF system design, using actual measurement data, and determine the specific cell site locations and configurations.

A.3 Preliminary Design Phase

Figure A–2 shows a flow diagram of the preliminary design process. Before the preliminary design process begins, various inputs must be provided. Some of the important inputs include the coverage boundary map and projected traffic, usually prepared by the service provider, and link budget and coverage/capacity testing requirements, provided by the system manufacturer. The link budget specifies such items as vehicle and building penetration loss, antenna gain, and the fade margin. Other inputs include the requirements and specifications for

- voice quality (FER),
- antenna types and heights,
- probability of service, and
- traffic density and distribution.

The rest of this section describes the major activities performed in each step of the preliminary design process.

A.3.1   Step 1. Project Plan and Requirements Review

The first step of the RF design process is to hold a series of meetings in which the service provider and system manufacturer

- create the project plan and deliverable time frames,
- confirm the design requirements,
- review the inputs needed and the content of those received, and
- determine and understand any ground rules and constraints.

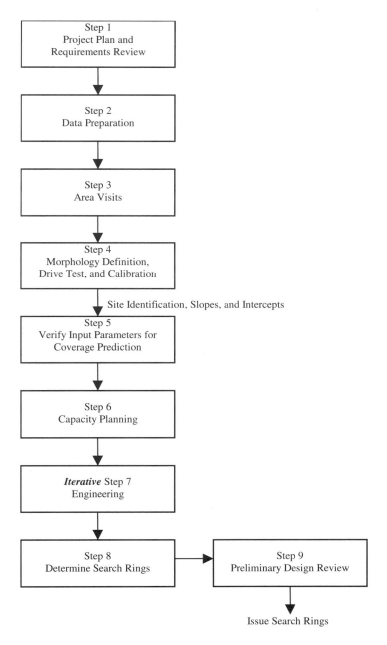

Figure A–2 Preliminary Design Process Flow

These meetings include identification and documentation of any changes to the requirements. All the parties involved should clearly understand the impact of those changes. These meetings will result in a documented understanding between the service provider and system manufacturer of the requirements and the finalized link budget that will be used for the design.

A.3.2 Step 2. Data Preparation

During the preparation step, the designer will gather information and data required to begin the design. This will include determining the availability of the required data and information and initiating a request for data where needed. The preparation step requires data for three major items.

1. Terrain—Digital representation of the ground elevation is needed for use in the RF coverage prediction tool such as CE4. The digital terrain data should be derived from an accurate source such as satellite imagery or 1:24,000 scale topographic maps. Data bin size of 100 m is generally sufficient; however, 30 m or smaller bins may be useful in regions of extreme elevation variation. Decreasing the bin size increases the number of RF tool calculations required for a given region.
2. Maps—Maps showing highways, streets, political boundaries, state boundaries, county boundaries, etc., are needed. Digital representation of this vector information is also needed in the RF tool. If not readily available, the maps will need to be digitized.
3. Boundaries—The service provider must provide service area boundaries. Marked areas will designate agreed-to areas of noncoverage within those boundaries. If boundaries are not provided in the same digital format as that for the maps, the data should be converted so that it can be used in a software tool for coverage prediction, such as CE4. In addition, when required by the service provider, service class boundaries should be specified and location-specific signal-strength requirements should be indicated.

A.3.3 Step 3. Area Visits

The area visits determine which morphology class each area falls into and initiate identification of candidate sites. It is very important for the RF design engineer to observe the area and develop a sense of environment for the PCS network being designed. During the visit, the RF engineer shall gather qualitative and quantitative data about the environment that include the following:

- terrain,
- trees (type, height, density),
- average heights of buildings in area (number of stories),
- building spacing,
- building locations,
- landmarks,
- airport and heliport locations,
- *difficult* areas where cell sites cannot be placed, and
- locations of existing cell sites, including those owned by competing service providers.

Quantifiable data includes photographs and location coordinates in latitude and longitude. Latitude and longitude coordinates may be carefully pulled from topographic maps or obtained using a Global Positioning System (GPS) receiver. Each area should be surveyed for preferred or desired locations of cell sites and potential *anchor* cells. (These are the most important, highest-priority search rings and are based on the morphology of the market and traffic capacity.) Photographs and coordinates for those locations should be collected. Each site/area visited should be documented as a package, including photographs.

A.3.4 Step 4. Morphology Definition, Drive Test, and Calibration

The geographic area of the design needs to be classified geomorphologically into like regions with the same morphology. The primary goal of this effort is the establishment of a set of region categorizations that can be used to accurately estimate path loss. In addition, coverage objectives for technical or contractual reasons may be specified on a morphology-by-morphology basis.

The classification process can be divided into two phases. The first phase establishes general *macro* morphologies (such as urban or rural). The second phase establishes *sub* or *micro* categories.

The first phase of this process exploits prior design experience which indicates that observable area aspects, such as terrain (flat/hilly/water), building density/height, road density and width, foliage, and drive path loss. Accordingly, this procedure primarily relies upon data and/or observations not obtained from RF signal strength measurements. When each morphology class has been mapped, drive testing of each class can be executed to obtain sufficient data for sub-classification. The drive-test route should be planned to ensure uniform data collection throughout the test area for each morphology class. This will ensure that the results are not biased to a particular area or morphology subclass. (Areas such as elevated roadways, cutout roads, tunnels, etc., would cause spikes in the data and should be carefully used.)

The second phase of sub-classification is primarily based on analysis that groups cells into sets, indicating comparable RF signal path loss. RF signal strength measurements collected during drive tests primarily drive this procedure, but area observation is also a necessary component. After eliminating any outliers, the RF design engineer integrates the measurements using the RF coverage prediction software (CE4). The software typically performs a standard linear regression on drive-test data. The path loss slope and one-mile intercept are estimated for each cell/area tested. Using this approach, the RF design engineer groups the results (that is, the intercepts and the slopes) of multiple tests within the same morphology class into any obvious sub-classes by comparing numerical values and by correlating these values with area observations, (for example, terrain and clutter causing any radio obstructions). This process should focus on groupings for the intercepts as these values primarily determine the path loss characteristics for a cell. Grouping requires some judgment and can be accomplished in a variety of ways.

Note that the procedures followed in each phase are not necessarily mutually exclusive, for example, a rural classification may be clarified and/or finalized by examining path loss data for the cell in question and comparing the results to those obtained for other rural cells. Similarly, a sub-class decision may be better defined by comparing visual observations of the cell in question with those obtained from others of the same sub-class.

It is important to keep in mind that *both* the link budget and the performance requirements may vary on a per-morphology basis.

A.3.5 Step 5. Verify Input Parameters for Coverage Prediction

Use of a software tool such as CE4 can predict the RF coverage. In running such a tool, the CDMA parameters from Step 1 (link budget parameters) and the slope and intercept values gained from drive test and calibration need to be supplied. Once they are verified, all of the input parameters should be documented.

A.3.6 Step 6. Capacity Planning

The RF design must satisfy the market capacity and coverage objectives. The design process first ensures the latter. After meeting coverage objectives, use traffic information to add any additional cells that may be required to address local capacity shortfalls.

Traffic data in the form of demand per area (for example, Erlangs/km^2) assesses whether capacity requirements have been met. Service provider traffic projections are often the means for obtaining this data. Available data can range from a simple estimate of the total offered traffic within the market (that is, the total Erlangs over the market area) to quite accurate, localized estimates of traffic demand. The latter is likely to occur when the design is to be overlaid on an existing wireless system (for example, AMPS); in this case, traffic demand within each sector of the underlying system is already known.

In some cases in which known or estimated information about the total traffic within the market is available, the traffic demand per area is computed by dividing the total by the service area. To meet capacity demands, this ratio must be less than or equal to the ratio of the offered capacity within a cell to its area footprint. Violation of this rule requires the addition of more cells or carriers. In this scenario, the market has become traffic-driven, rather than coverage-driven, since these additional cells do not increase coverage but only add capacity

Use of additional carriers achieves capacity relief at the cost of some additional design complexity. For example, this solution requires clearance of additional spectrum. For an existing underlying system, clearance can result in additional blocking and consequent loss of revenue. In addition, small *localized* additions of a second carrier (for example, a single-cell hot-spot) may require the placement of the second carrier in surrounding cells to facilitate handoff from carrier 2 to carrier 1 for mobiles traveling out of the hot-spot area. Finally, the computation of multi-carrier capacity across an extensive area may not be straightforward in a technology like CDMA where the service provider may elect to operate both carriers independently to avoid performing hard (carrier 1 to carrier 2) handoffs for mobiles traversing the area. Accordingly, a viable first-order approach for second-carrier designs places the second carrier where needed and assumes that total capacity can be computed by assuming a fixed number of Erlangs per carrier. This process establishes a lower bound on cell count as well as on the multi-carrier capacity.

A.3.7 Step 7. (Iterative) Engineering

Now that the design region is understood from morphology and capacity perspectives, the iterative design engineering begins, and the engineer will perform the following tasks.

1. Evaluate the pre-qualified site candidates, if any, the service provider selected. Identify those sites that can be used for the area.
2. Define *anchor* cells (which include traffic capacity, RF trouble spot, and other *must have* cell locations). These cells dictate the overall RF network design. They, in a large sense, determine the rest of the search rings. Generate an initial cell-site layout, starting with anchor cells and using the preferred/desired locations and the pre-qualified site candidates.
3. Identify new sites needed for coverage or capacity. Concentrate on coverage first since coverage holes are undesirable.
4. In the RF tool, input the pre-qualified sites that were identified as usable and the new sites needed for coverage and capacity.
5. Run the propagation model to obtain initial area-coverage results.
6. If there are holes or excessive overlap in coverage, revise the cell layout by moving cell-site locations to eliminate the holes or overlap. Check the revisions by running the propagation model again to check coverage.
7. Iterate Steps 3-6 until a preliminary design satisfies the requirements.

A.3.8 Step 8. Determine Search Rings

Search rings define the areas where a need for antenna placement has been determined. Search rings are not precise cell-site locations.

1. Determine the search-ring radius based on the area (morphology) characteristics. The drive-test data reveals a nominal radius for each morphology. The RF engineer determines search rings, keeping in mind that there are two types of cells: those driven by traffic capacity and those driven by coverage.

    A rule of thumb is to use one-fourth of the cell-site radius as a search-ring radius. However, it may be more accurate to consider potential revenue and the voice quality desired. For example, in areas with high traffic, the search-ring radius would be smaller since you are concerned both with traffic capacity and voice quality. In farmland areas, the search-ring radius may be larger since the main concern is coverage.

2. Identify *anchor* search rings first.
3. Identify additional search rings, starting in the area immediately adjacent to the *anchor* search rings and keeping in mind how each search ring affects the others.

A.3.9 Step 9. Preliminary Design Review

The final step of the preliminary design is the review and verification of the design that has been completed according to the steps explained thus far. Key inputs for the preliminary RF design review include the following:

- service boundary map,
- drive-test data with slope, intercept, standard deviation of error, standard deviation of slope fading, and standard deviation of intercept,
- table of slopes and intercepts used by morphology class/sub-class,
- preliminary total cell count, and
- preliminary cell-site locations and search-ring map.

Following a successful design review, the RF design engineer may release the search area rings to the site acquisition team. If possible, the engineer will work with site acquisition to secure locations in *anchor* search rings prior to issuing subsequent search rings.

A.4 Final Design Phase

Figure A–3 shows the steps involved in the final design phase. Each of these steps will be described in the following sections.

A.4.1 Step 1. Candidate-Site Selection

After the site acquisition team has identified candidates for each site location/search ring, the RF engineer will review the candidates according to the following criteria and determine if the site is an acceptable candidate for a RF team site visit:

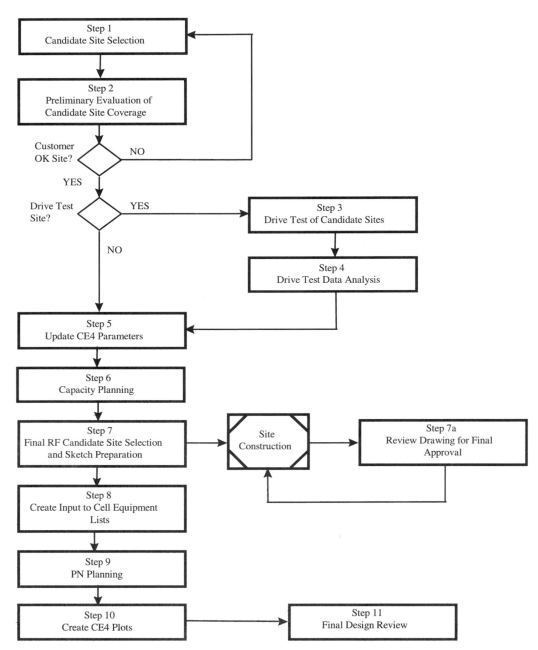

Figure A-3 Final Design Process Flow

- cell site location latitude/longitude (is it within the search ring?),
- antenna height limitation (antenna height clearance, as a rule of thumb, is fifteen feet above clutter),
- zoning restrictions,
- surrounding characteristics (use pictures, video, maps, etc.), and
- coverage as indicated by a software tool (for example, CE4) used to check the coverage objective and to find sites that provide optimal coverage (three-second terrain data or better (where available) should be used throughout final design).

If the candidate site is not acceptable, the site evaluation team issues a request for new candidates. Once all potential candidates have been exhausted, a new search ring will be issued.

A.4.2 Step 2. Preliminary Evaluation of Candidate-Site Coverage

During the candidate-site visit, the site evaluation team performs the preliminary evaluation. This team consists of RF engineering, site acquisition, and implementation personnel. The purpose of these site visits is to gather information on the candidate sites so that design and implementation decisions can be made.

During the site visit, photographs and GPS coordinates of the site location should be obtained. If the candidate site is a roof top, rather than a tower, building height from building engineering drawings should be recorded and pictures from the antenna-mount position (360 degrees) should be taken. Also, the proposed antenna installation should be documented. Sketches showing cell placement and antenna mounting should be made. For a preliminary evaluation of the cell coverage, note the following:

- the antenna's view (line of sight or shadow) to major highways and other target areas,
- other search rings and candidates to see how compatible they are, how they overlap, and how they might cover each other's holes,
- other possible candidates, and
- characteristics of area, especially building or tree height/spacing and building types (commercial, residential, or industrial).

The service provider should review the results of the site visit and make a business determination of whether or not the candidate is acceptable. For acceptable candidates, an RF determination should be made regarding drive testing of the candidates.

A.4.3 Step 3. Drive Test of Candidate Sites

The percentage of sites that require drive testing will depend on engineering judgment and on the severity of the terrain. When performing the drive test, the engineer should take the following steps to limit uncertainty.

- Classify morphology visually at the site and by using aerial photography or maps.
- Uniformly drive around the planned service area of the site to avoid weighting biases and use spatial data collection mode (versus time-based collection mode) whenever possible.
- Instruct the drive-test team to drive until they lose the signal.
- Check the transmitter and receiver calibration and the correct transmit power and VSWR at the start and end of the test using a Watt-meter.
- Choose the highest possible transmit power.
- Use low-gain, omnidirectional transmit antennas and positioning at design heights at least fifteen feet above the surrounding clutter.
- Apply engineering judgment to partition the drive-test data set when drive routes exhibit very different morphology characteristics, for example, straight highway shrouded by dense foliage, elevated highways above urban areas, shadowing due to unavoidable clutter, etc.

A.4.4 Step 4. Drive-Test Data Analysis

The RF engineer should review the drive-test data for verification purposes and to minimize uncertainty and perform the following tasks.

- Prune drive-test data prior to a measurement integration to eliminate non-linearities due to power measurements exceeding the compression point (non-linear operation) of the receiver (typically greater than -60 dBm).
- Eliminate near- and far-data points (for example, less than 1 Km and data within 5 dB of the receiver noise floor).
- Run a measurement integration (using, for example, the CE4 measurement integration feature) and analyze the result with regard to the previous division of the cell into morphology classes. Find the optimal division of the cell into morphology classes (arcs) by minimizing the standard deviation of the error between measured and predicted data.

- Derive the slopes and intercepts for each significant radial segment of the coverage area by analyzing drive-test information.
  - **a)** Examine the drive-test data plots and determine the angular sector boundaries of the propagation. (For example, look at the site from a sector or radial segment viewpoint.) If those are different morphology classes (water or close to water, highway, forest, etc.), then group these separately from other path-loss groups. Another indicator of a different path-loss group is the error in dB between the predicted and the measured value. As a rule of thumb, if the error is larger than 12 dB, try separating the groups into radial segments with different slopes and intercepts.
  - **b)** Run the CDMA Forward Link Strongest Pilot System Wide Server analysis (in CE4).
  - **c)** Repeat the correction cycles on an arc-by-arc basis. After a second iteration, check the inputs.
  - **d)** Recognize areas of similar morphology in subsequent sites and begin predicting the behavior of future drive-test segments. Be sure to construct the drive-test routes and parse the data to facilitate subsequent analysis.
- Compute lognormal standard deviation to get a feel for the validity of the data set. If the standard deviation is much larger than what is required, apply a margin if the data set is deemed valid. The margin is applied to compensate for the difference in assumed lognormal standard deviation in the link budget of 8 dB. Use 8 dB as the default standard unless otherwise specified in the link budget for the particular project.
- Apply engineering judgment to define special morphology classes associated with each partitioned data set; exclude partitioned data sets from statistical analysis where it makes sense.

A.4.5 Step 5. Update Parameters Needed for Coverage Prediction

After the drive test has been completed, the parameters (for example, CE4 input parameters) need to be updated to reflect the new information.

There are two aspects to this step. The first is to update the cell parameters in the RF tool. For the cells that have been drive tested, the slope and intercept are as determined by the measurement integration process. Adjustments are also made for long cable runs or antennas not covered by the link budget. For cells that were not drive tested, the adjustments to use result from the combination of the slope and intercept for that morphology and site specific factors such as cable losses and antenna gains.

The second aspect to this involves either firming up or revisiting the other parameters used for the CDMA analyses. It is possible that over the course of the design some items such as *T_ADD*, *T_DROP*, or voice activity factor may be changed, and those changes need to be reflected in running a coverage prediction software.

A.4.6 Step 6. Capacity Planning/Traffic Studies

Refer to Section A.3.6 for capacity planning and traffic studies.

A.4.7 Step 7. Final RF Candidate Site Selection and Sketch Preparation

The RF engineer's final selection of a (single) candidate site (out of many) is a two-part process, each of which is based on the consideration of the candidate on its own merits and demerits and the consideration of the candidate in the context of the current view of its surrounding neighbor sites.

Upon completion of these assessments, the checkpoints defined in A.4.7.1 and A.4.7.2 are documented. Then the final RF candidate for the site is selected. It should be noted that at any time between this point (final candidate selection complete) and the actual construction of the site, something unforeseen could cause this site to be lost or require it to be reassessed as part of the larger system. If this occurs, the RF engineer will have to return to Step 1 or Step 2 of the final design process and begin again to determine the best candidate for this site area and possibly one or more of the surrounding site areas.

A.4.7.1 Consideration of the Candidate on its own Merits and Demerits

This process builds on the observations and notes made during the initial candidate-site visit and other collected data and analyses such as drive testing, isolations, and intermodulation studies. Ideally, all constraints on the candidate were known or determined at the time of the initial candidate visit; however, if something significant changes or the observations were insufficient, another site visit should occur before completion of the final site selection.

Some of the key questions to be asked in assessing the candidate on its own merits are as follows.

- Are this site area and its candidate(s) part of the official design area?
- Does the candidate satisfy the coverage objectives of this site?
- Are AM broadcasting tower related risks identified and evaluated?

- Are FAA requirements satisfied?
- Does sufficient isolation exist between the PCS antennas and other antenna systems at the candidate location?
- Can the GPS antenna be properly positioned to ensure signal reception?
- Given the desired (and permitted) antenna locations and the base station equipment placement, are there any link budget exceptions anticipated such as a long coaxial cable run.
- Is the desired antenna model(s) available?

The choice among multiple candidates cannot yet occur. Each viable candidate must be assessed in the context of its neighboring sites.

A.4.7.2 *Consideration of the Candidate in the Context of its Neighbors*

Except for the rare instance, each cell site will have one or more neighbors. The design of one site can have bearing on the design of neighboring sites, and for that reason candidate assessment in a vacuum is insufficient. Each contending candidate needs to be assessed in the context of the current view of the neighboring site. As the design for the neighbors may not yet be complete, the current view is simply the RF engineer's opinion of the neighbors' most likely configuration. The opinion amounts to an assumed location (latitude and longitude), height, antenna model, and orientation and should be documented. With the view established, the RF engineer judges the candidate in context and on the following checkpoints:

- satisfaction of the coverage objectives of this site in the vicinity of its neighbors,
- achievement of anticipated handoff zones, and
- minimization of the potential for internal system interference.

A.4.7.3 *Final Candidate Selection*

Once all candidates for a site have been fully considered, the one best meeting the objectives of the site in the context of the overall design is selected as the final candidate.

A.4.7.4 *Sketch Preparation*

Following the above-described candidate assessments, final candidate selection, and documentation, it is necessary for the RF engineer to create a site configuration form, which includes a sketch and will be given to the construction team to use in the creation of various drawing packages. The site configuration form needs to include all relevant physical information about the site, including that listed below, plus a sketch with reference points indicating the orientation of the antennas and the mounting structure to be employed:

- site identifier, candidate identifier, and candidate name,
- RF engineer name and date of preparation,
- reviewer name, date, and signature,
- structure type and configuration (omni or sectored),
- per-sector specifics: number of antennas, antenna manufacturer, antenna model and dimensions, horizontal and vertical beamwidths, azimuth with reference to either true or magnetic north, horizontal separation, estimated coaxial cable run length, assumed jumper configurations, coaxial cable type (for example, 7/8 inches), amount of mechanical downtilt, centerline (or other antenna reference point) height relative to ground level or other site reference point, calculated total value of the anticipated coax, connector and jumper losses combined,
- GPS antenna specifics: GPS antenna model, coax type, estimated cable run and loss, GPS antenna height above a stated reference, and
- space for relevant or explanatory comments.

The sketch portion of this form should indicate graphically much of the positioning information listed above plus other relevant data such as building dimensions, cell equipment location, a reference for north, etc. The sketch should also specify the type of mount, for example, sled versus pole mount, to back of parapet wall, etc., and the exact location of the GPS antenna.

A.4.7.5 Step 7a. Review Drawing for Final Approval

The RF engineer then reviews the site acquisition form and drawing for any discrepancy with regard to the site configuration form before final approval. Major items to be reviewed are as follows:

- antenna locations, orientations,
- antenna type, height, spacing,
- antenna mounting types,
- antenna tilting,
- cable length, type, and size, and
- pictures of view from antenna.

A.4.8 Step 8. Create Input to Cell Equipment Lists

Before the cell sites are ordered, RF design input to the final cell equipment list should be given to construction or program management. The RF input to the cell equipment list should at a minimum contain the items listed below:

- number of carriers in cell,
- cable type and length,
- cell TX and RX frequency, and
- antenna type and model number.

A.4.9 Step 9. PN Planning

See Chapter 4.

A.4.10 Step 10. Create Coverage Prediction Plots

Once the RF design is complete, a set of the coverage prediction plots needs to be generated. The plots listed below are typically generated using a tool such as CE4:

- system-wide strongest signal strength plot,
- system-wide strongest serving pilot E_c/I_o plot,
- system-wide soft handoff boundaries plot,
- system-wide balanced reverse-link coverage plot, and
- traffic analysis plot.

The plots generated for a particular market should all be the same size and scale, preferably 1:100,000 or 1:150,000. If the system is so large that it does not fit on a single plot, then all plots should consistently show the system broken down into smaller pieces.

A.4.11 Step 11. Design Review

The last step of the RF design process is to review and verify the final design in which coverage requirements, service boundaries, path loss slopes and intercepts used by morphology classification, drive-test data along with its analysis results, and traffic analysis are included. Once everything is verified, final output is generated. Included in the final output are the following:

- exact location of cell sites and antenna locations,
- PN offset assignments and neighbor lists,
- coverage prediction plots,
- RF engineering site sketch, and
- drive-test data.

APPENDIX B

Outline of RF Optimization Procedures

The RF optimization procedure defines field tests intended to tune all aspects of the CDMA air-interface network performance. The optimization procedure consists of two phases: cluster testing and complete system-wide optimization. In other words, the initial pass of the RF optimization will be performed as a part of the cluster tests; the second, more detailed, tuning phase will occur after completion of all cluster tests, which occur after activation of all cell sites in the CDMA network. The primary reason for breaking the optimization work into two phases is to reduce the time and resources required to complete the cluster test cycles.

The most significant objectives of the optimization testing are the following:

1. to ensure the achievement of acceptable coverage for the pilot, paging, synchronization, access, and traffic channels,
2. to minimize the number of dropped calls, missed pages, and failed access attempts,
3. to control the overall percentages of one-, two-, and three-way soft/softer handoff, and
4. to provide reliable hard handoffs for CDMA-to-AMPS or CDMA f1-to-f2.

During the RF optimization process, CDMA parameters will be adjusted using simulated traffic loading due to the extreme cost and logistical obstacles associated with employing live traffic. In all cases, forward-link loading can be implemented using a noise simulator that simulates the forward-link orthogonal channels. Achievement of reverse-link loading is through the use of an attenuator and circulator at the mobile.

B.1 Cluster Testing

Cluster testing consists of a series of procedures to be performed on geographical groupings of approximately nineteen cells each; roughly, the clusters are selected to provide a center cell with two rings of surrounding cells. The number of cells in each cluster is relatively large to provide enough forward-link interference to generate realistic handoff boundaries in the vicinity of the center cell and the first ring of cells; a cluster of fewer cells would provide acceptable results over too small a geographic area. Provision of approximately one tier of cell overlap between each cluster and the next affords continuity across the boundaries. The goal is to complete all tests for a given cell cluster, while minimizing the utilization of test equipment, personnel, and time. For this reason, the intent of cluster testing is to coarsely tune basic CDMA parameters and to identify, categorize, and catalog coverage problem areas. No attempt will be made to resolve complex time-intensive performance problems during the cluster test phase; such location-specific, detailed refinements will be deferred until the system-wide optimization phase.

B.1.1 Spectrum Monitoring

The first preliminary step in the cluster testing involves monitoring uplink and downlink interference in the CDMA band to verify that the spectrum is clear enough for CDMA operation. For the CDMA system to properly function, the spectrum must be cleared in a sufficient guard band and guard zone. Even the intermittent presence of strong in-band interference can significantly degrade radio performance for the CDMA system. In extreme cases, it can be a time-consuming, difficult task to identify and mitigate the sources of external interference. These can be, for example, microwave data transmissions, externally generated intermodulation products, and wideband noise from arc welders and other machinery, etc.; therefore, it is important to begin spectrum monitoring as early as possible. These spectrum-monitoring tests also provide a baseline data set of measured background interference levels that can be used to optimize reverse overload control thresholds and jammer detection algorithms for specific environmental conditions.

B.1.2 Basic Call Processing Test

The next stage of the cluster testing prior to optimization is to exercise basic call-processing functions, including origination, termination, and handoff, to assure that these basic telephony capabilities are operational. Quick measurements are then made of CDMA signal levels to verify that each cell site is transmitting the appropriate power levels. The intent of these basic functional tests is to detect hardware, software, configuration, and translation errors for each cell site in the cluster prior to drive testing. This sector-testing phase will involve driving in the coverage area of each sector to assure that installation has been correctly completed. At this stage, it is common to detect bad coaxial cables/connectors, mis-oriented antennas, and other similar defects.

B.1.3 Unloaded Pilot Survey

After basic cell operation has been verified, surveys of forward-link pilot channel coverage are performed with light traffic load on the system. During the unloaded survey measurements, all cells in the cluster are simultaneously transmitting forward-link overhead channels (for example, pilot, sync, and paging), with only a single Markov call (that is, a simulated voice call) active. The system manufacturer and the wireless service provider jointly select the drive routes used in the measurements. In general, the drive routes will include major freeways and roadways within the designed coverage area of the cluster where high levels of wireless traffic are to be expected. Drive routes may also be selected to explore weak coverage areas and regions with multiple serving cells as predicted by propagation modeling software (for example, CE4) or based on knowledge of the surrounding terrain topography.

The unloaded pilot survey results identify coverage holes, handoff regions, and multiple pilot coverage areas. The pilot survey information highlights fundamental flaws in the RF design of the cluster under best-case, lightly loaded conditions. The pilot survey provides coverage maps for each sector in the cluster; these coverage maps are used during the optimization phase to adjust system parameters. Finally, measuring the pilot levels without load serves as a baseline for comparison with measurements from subsequent cluster tests under loaded conditions. Characteristics of cell shrinkage can be compared under the extremes of light and heavy traffic load.

During the unloaded coverage tests, two iterative passes of optimization are conducted:

1. the first pass optimization entails correction of neighbor lists and adjustments to the fundamental RF environment (transmit power, antenna azimuth, height, downtilt, antenna type), and
2. the second pass at optimization involves fine adjustment of handoff thresholds, search windows, overhead channel transmit powers, and access parameters.

The focus of these two iterative optimization passes is on resolving problems observed by the field teams from the coverage plots and from analysis of dropped-call mechanisms.

B.1.4 Loaded Coverage Test

The final measurements to be performed as a part of the cluster testing are coverage drive runs conducted under loaded conditions. Drive routes for the loaded coverage testing will be exactly the same routes as those used for the unloaded coverage surveys. The objectives are to provide coarse system tuning and to identify, categorize, and catalog coverage deficiencies so the more difficult problems can be resolved during later system-wide optimization tests. During the loaded testing, both first-pass and second-pass tuning parameters are adjusted to fix problems observed by the field teams. At the conclusion of the loaded coverage tests, a performance validation procedure is conducted to measure system performance against the cluster-test exit criteria.

More often than not, the actual cluster testing should be kept as quick and simple as possible. This can be done by deferring much of the detailed system tuning to the subsequent system-wide optimization phase. Once the coverage deficiencies have been identified for a particular cluster, if a specific problem cannot be resolved in approximately one-half hour, then the field team will note the situation and proceed with the drive testing.

B.2 System-Wide Optimization

After all clusters in the CDMA network have been tested, system-wide optimization will begin with all cells activated. Optimization teams will drive-test each of the problem areas identified during the cluster testing, using the same test conditions under which the problem was previously observed. Iterative-tuning procedures will be used to fix coverage problems by adjusting transmit powers and neighbor list entries. Extreme situations may require modification of handoff thresholds, search window sizes, or other low-level tuning parameters. If the field team cannot resolve coverage problems in one hour, then the team will flag the problem area for further investigation by other RF support personnel.

After the site team has made attempts to resolve the individual coverage problem areas, the system-wide optimization will proceed to the final phase. The final optimization step will be a comprehensive drive test. This will cover the major highways and primary roads in the defined coverage area for the CDMA network. During the system-wide drive run, simulated loading will be used to model traffic on the network. Performance data will be collected as a small number of active CDMA subscriber units traverse the system-wide drive route. Statistics will be collected to characterize pilot, paging, traffic, and access channel coverage over the entire drive route. Specific problem areas identified by the system-wide drive run will be addressed on a case-by-case basis after the entire drive has been completed. Comprehensive statistics from the system-wide drive will be used to assess the overall performance quality of the network, including origination failure rate, dropped call rates, handoff probabilities, and frame error statistics.

At the conclusion of the comprehensive, system-wide drive phase, the RF optimization procedure will be considered complete and the CDMA network ready for live traffic testing and market trials leading into commercial service. Once significant loading with live traffic is present on the CDMA network, additional tuning of the system parameters will be required to accommodate uneven traffic conditions (for example, traffic hot spots, unusual traffic patterns, etc.) and other dynamic effects that cannot be easily predicted or modeled with simulated traffic loading.

APPENDIX C

RF Coverage Prediction with CE4

C.1 Overview of the CE4 Cellular Engineering Tool

Cellular Engineering 4 (CE4) is an RF planning tool that RF engineers use to generate predictions of the performance of the RF link of the system. As inputs, the tool takes information about the terrain (elevation and surface items such as trees and buildings) and the location and physical properties of the transmitting antennas (height, antenna type, azimuth, tilt). CE4 will predict the strength of the signals coming from each of the antennas and combine these predictions to produce predictions of coverage area, interference levels, SIR, etc. The current outputs of CE4 are plots that display the results of the requested analyses.

CE4 includes RF propagation analysis, interference prediction, and utilities that help create and manage system designs for both mobile and fixed wireless systems in multiple operating environments.

C.1.1 Tool Input and Scenario Setup

CE4 takes information about the terrain as input for the main predictions. The system area is broken up into discrete areas that are called *bin*s. A bin is the smallest unit of measure in CE4. Bin sizes are based on degrees of arc and can range from one second to sixty seconds per bin with typical values of three or six seconds. CE4 assigns an elevation level to the terrain for each bin in the service area. It is also possible to import information about what is on the surface of the terrain, for example, trees, open area, water, buildings, etc. The RF engineer enters this land-use data, or *clutter* as it is also known, on a per-bin basis so that each bin can contain only one type of clutter. The RF engineer is then able to define the effect (loss or gain in the signal strength) that the clutter will have on the RF prediction.

The scenario is a single version of a system design in CE4. It is the container into which all of the elements of a system are placed, including the terrain elevation and clutter, the cell sites, antennas, MSCs, etc. To generate a new design, the user starts by creating a new scenario. As part of this creation, the user defines the system type (CDMA, FM, fixed wireless, etc.), the spectrum that the system will use (Cellular A-band, PCS B-band, etc.), and the terrain to be used, and indicates other options (such as the preference between feet and meters). Once this is defined, the user can open the scenario and start the actual design work.

C.1.2 Cell-Site Placement and Provisioning

Once the scenario has been created and opened, the RF engineer can begin to define the system. The engineer starts by defining one or more MSCs. Since all cell sites must be *attached* to an MSC, at least one MSC must be defined in the scenario to be able to start defining cell sites.

Once there is at least one MSC, the RF engineer can proceed to define cell sites. The cell-site definition process requires that the engineer choose a location for the cell site (this is normally done by choosing a location on the display) and then defining *emitters* that are to be used in the cell site. (An emitter can be thought of loosely as a set of radios plus an amplifier plus an antenna.) As part of this definition, all of the parameters needed for the particular system type need to be defined, including the antenna type, height, ERP, tilt, and azimuth.

To increase the productivity, all parameters are given default values (which the engineer may change) and it is possible to copy a cell site and paste it into many different locations, thereby making cell-site replication quick. Once all of the cell sites have been defined, the engineer is ready to perform analyses.

C.2 Analysis Features Available for the Demo Version of CE4

The CE4 software tool that is included on the CD-ROM that accompanies this book is a demonstration version with limited capabilities. This section describes some of the key features available for this demo version of CE4. The features excluded from this demo version, but are available for the full version, are described in Section C.4.

C.2.1 Signal Strength Analysis

CE4 performs the signal strength analysis in the same manner, regardless of system type. When the engineer requests that an analysis be performed, CE4 will first perform a signal strength prediction for all of the emitters that are *turned on* (the user can select which of the emitters to include in an analysis). CE4 performs the signal strength analysis for each of the emitters one at a time. The output of the analysis is a file for each emitter; this file contains the predicted signal strength for each of the bins that the emitter reaches. (There is a finite calculation distance.) The file also contains information that allows the software to determine which of the bins are included in the file, that is, the calculation area.

A slope and intercept model that accounts for the effects of the terrain as well as the clutter on top of the terrain forms the basis of the signal strength prediction.

C.2.2 CDMA Analysis

Some of the analyses that are available for CDMA systems are described below.

<u>Forward Link Analyses</u>

1. Strongest Pilot E_c/I_o provides a prediction of the E_c/I_o for the strongest pilot on a per-bin basis.
2. Pilot Overlap provides an estimate of the number of pilots above a user-defined E_c/I_o threshold that exists in each bin.
3. Strongest Traffic E_b/I_o provides a prediction of the E_b/I_o for the strongest traffic channel (emitter with the strongest traffic signal) on a per-bin basis.
4. Nominal Soft Handoff Boundaries provides a prediction of where the mobiles are expected to be in soft handoff and shows where each type of soft handoff is predicted to occur.
5. Signal Strength indicates the signal strength for each of the logical channels in the CDMA carrier (pilot, paging, sync, traffic).
6. Single Server Analyses predict the signal level and the signal-to-interference ratio of the logical channels.

Reverse Link Analyses

1. Coverage plot for the reverse link predicts where the mobile will have sufficient power to reach a base station with the required E_b/I_o.

Balanced Links

1. Balanced Coverage analyses predict where there will be coverage in both the forward and reverse directions.
2. Unbalanced Coverage predicts where there will be coverage in the forward direction but not the reverse or vice versa.
3. Balance Traffic analysis predicts the traffic load on each emitter based on the Balanced Coverage analysis (traffic assigned to an emitter only if the bin has coverage in both directions).

C.3 System Requirements and Installation

CE4 is designed to run on personal computer systems. We recommend the following to run CE4 properly:

- computer: IBM-compatible desktops or laptops (CPU speed at 133MHz or faster),
- operating system: *Windows* 95 or *Windows NT*[1],
- memory: 32 megabytes,
- swap file: 50 megabytes,
- disk and file access: 32-bit, if system allows it,
- system resources: To improve performance, close down other applications before running CE4. (including removing wallpaper and screen savers), and
- energy-saver hardware/software: Disable energy savers if you are running an analysis that will run long enough to activate the energy-saving program.

See also the "Hardware Configuration Details" section in the CE4 software on-line Help.

---

1. After you install CE4, create a shortcut icon to CE4 MacroPlanner (you should already have ceauto.exe in ...\ce4\bin) and check the "Run in Separate Memory" box. This allows running more than a single session of CE4 without rebooting *Windows NT*.

To install the software, follow the procedure described below.

- Place the CD-ROM into the drive.
- Using file manager (or explorer), navigate to the CD-ROM drive.
- Double-click on setup.exe in the "disk1" directory of the CD-ROM.
- When asked, fill in company name and your name. Note that any value is allowed, but these fields must not be blank. Enter the value "1" for the serial number. Click OK.
- At the select application dialog, leave the default (CE4 MacroPlanner). The other options are not included in the demo version. Click OK.
- Choose a directory in which to install CE4. After entering the path to the directory, click OK.
- CE4 will now install to the chosen directory with no further input from the user.
- To start CE4, double-click on the CE4 icon.

C.4 Limitations of the Demo Version of CE4

This section explains some of the features that are not executable in the demo version of CE4. The restrictions imposed include the following.

- Export/Import of scenario data and supporting files is not allowed.
- Scenario type is limited to CDMA. The other types are disabled.
- Terrain (only the demo file included) bin size is 25 seconds by 18.75 seconds (760m x 465m) to reduce computation time required to complete CDMA analysis.
- Measurement integration feature[2] is disabled.
- Printing is not allowed. (Screen dump is possible if one has an application program that supports it.)

---

2. By their nature, the propagation models used in all of the RF prediction programs are great simplifications of the real world. Because of this, it is necessary to be able to calibrate the model for the local conditions that exist. This is done in CE4 by means of measurement integration. Measured data for the area is imported into CE4 and used to adjust the slope and intercept values of the model so that the model more closely reflects the local conditions. This is done by fitting a curve to the difference between the measured and predicted data and extracting the required change in slope and intercept to minimize the error. CE4 performs this calibration process one emitter at a time.

C.5 Walk-Through of the CE4 Tool

In spite of the restrictions, one can still run the included CE4 software tool to understand how such a tool is used for the coverage prediction step of the RF design process described in Appendix A. In addition, one can also use it as a simulation tool to examine the effects of different values of CDMA parameters.

As an illustration, Figure C–1 (see inside front cover) shows soft handoff plots for a seven-cell cluster (assuming flat terrain and with cell sites that are apart by about six miles) for two different sets of T_ADD/T_DROP parameter values, along with all other parameter values shown in Table C–1. This section details the steps involved in getting such a plot. For more information on different menu options, one can refer to the online Help included in the software.

Start the CE4 Program

1. Double-click the CE4 icon located in the CE4 Program Group or on your *Windows* desktop. An entry for the Demo folder and any other folders that you create appear on the left side of the Scenario List.
2. Double-click Demo. The Demo folder contents appear on the right of the window, including demo scenario files[3] that are shipped with the CE4 software and any scenario files you create using the CE4 tools.

Create a New Scenario and Network Elements

1. In the Scenario List window, select [File|New Scenario…]. The [New Scenario] dialog box opens.
2. In the [New Scenario] dialog box,
    - select Demo for the Folder, and type in a Scenario name,
    - click on "CDMA a band" for the Spectrum in the [Attributes] box,
    - click on Terrain to open the [Select Terrain Directory] dialog box, choose DUMMY for Directory and click OK, and
    - press OK back in the [New Scenario] dialog.
3. In the Scenario List window, double-clicking the scenario name that was just created opens the scenario, and the Scenario window is displayed. Because you

---

3. The demo folder contains a demo CE4 scenario (collections of geographic and system data) entitled CDMA that will help you get started using CE4. The demo scenario contains sample terrain and antenna data files. To perform analyses on a real system, you would need to acquire actual terrain and antenna pattern data. Note that the demo version of CE4 does not support importation of these files.

have not yet created any network elements, only the blank scenario background appears in the display.

4. Create an MSC. Choose [Edit|New MSC], left-click mouse, and open the [New MSC] dialog box. Type in Number (for example, "1") and System ID (for example, "1"), and press OK. A small square, indicating an MSC, will be created where the mouse was clicked.

5. Create a new cell site.
   - Choose [Edit|New Cell Site]. Left-click at the location where a new cell is to be created, and open the [New Cell Site] dialog box.
   - Choose CDMA for the Emitter Type. Clicking on the box in the More column opens the [Change CDMA Emitter] dialog box, in which you can verify the parameter values shown in Table C–1.
   - Click OK in the [New Cell Site] dialog box. An omni cell will be created at the location where the mouse was clicked.

6. You can repeat the Step 5 for the other cells to be created (to form a seven-cell cluster, as shown in Figure C–1 on the inside of the front cover), or you can use the copy-and-paste method. For the latter method, choose [Edit|Copy Cell], and click inside the cell just created. Then choose [Edit|Paste Cell], move the mouse to the location for a second cell and left click, press OK in the [Paste Cell Site] dialog box, and repeat this for the other cells.

Run the CDMA Analysis

1. Select Cells for Display or Analysis.
   - Once all of the network elements have been created, select the arrow icon (the first icon) from the toolbar.
   - Left-click inside the cell symbol for each cell you want to select. The symbols for the selected cells change color. For the default settings, red indicates the cell is *on* and blue indicates it is *off*. You can also select all of the elements using [Select|Select All].
   - If cell-site parameters need to be examined or changed, right-click the cell site for which the parameter is to be changed. [Change Cell Site] dialog box opens. If the parameter to change is not found in the screen, click the More button, and then open the [Change CDMA Emitter] dialog box.

2. Specify CDMA Analysis Parameters.
   When you have opened a CDMA scenario and selected the cell(s) you want to include in the analysis, you are ready to specify the analysis parameters.

- In the Scenario Display window, select [Analysis|CDMA Analysis]. The [CDMA Analysis] dialog box opens.
- Select the link (Forward, Reverse, or Balanced) you want to analyze for the selected cell. Select "Forward Link."
- In the [Analysis Type] box, highlight the link characteristic you want to calculate first. Select "System-Wide Nominal Soft Handoff Boundaries."
- Note the maximum mobile ERP setting in the [Analysis Parameters] box. Specify all of the CDMA Analysis Parameters, if needed.

3. Run a CDMA Analysis:

   When you have opened a scenario, selected the cell(s) you want to include in the analysis, specified the analysis parameters, and specified the plot band options, you are ready to perform the analysis.
   - In the [CDMA Analysis] dialog box, press OK. CE4 performs the analysis on the selected cells according to the analysis parameters you selected. A [Progress] window opens to keep you apprised of how far the analysis has progressed. When the analysis is complete, the results are displayed.
   - To clear the analysis display from the [Scenario Display] window, press [Analysis|Clear All Analysis].
   - To restore an analysis display that has been cleared, press [Show Current Analysis]. You can customize the analysis results display using the [View|Options].

4. Change Cell Site Attributes:
   - After the CDMA Analysis with the default parameter value has been run, which can produce a plot such as Figure C–1a (see inside front cover), you can change parameter values (for example, *T_ADD* and *T_DROP*) for all of the cells by choosing [Edit|Change Selected Emitters...] and opening the [Emitter Parameters] dialog box. In this case, all of the highlighted (that is, *on*) cells are affected. In [Emitter Parameters] dialog box, check the "change" column of the parameters you want to change, click on the data column, and type in new values.
   - Press OK in [Emitter Parameters] dialog box. You can verify the changes you made in the [Change CDMA Emitter] dialog box, as described in Step 1.

Table C-1 CDMA CE4 Default Parameters*

System Parameters		CDMA Emitter Default Parameters	
Spreading Bandwidth (PN Chip Rate)	1.23 MHz	Average # of Users/Emitter	13
Voice Activity Factor (Forward)	0.48	Average Erlangs/Emitter	20
Voice Activity Factor (Reverse)	0.4	Pilot Pwr ERP	23.94 Watts
Base Station Noise Figure	5 dB	Sync Pwr ERP	2.39 Watts
Forward Traffic/Voice Data Rate	14400 bps	Paging Pwr ERP	8.42 Watts
Reverse Traffic/Voice Data Rate	14400 bps	Traffic Pwr ERP	10.84 Watts
Paging Channel Data Rate	9600 bps	Minimum Forward Traffic Eb/No	7 dB
Sync Channel Data Rate	1200 bps	Minimum Pilot Ec/Io	-15 dB
Reverse Propagation Loss Factor	0.00	Minimum Sync Eb/No	7 dB
Mobile Noise Figure	10.7 dB	Minimum Paging Eb/No	9 dB
Fwd Traffic Interference Reduction Factor	0.00	T_ADD	-12 dB
Fwd Traffic Fade-Soft Handoff Gain	0.00	T_DROP	-15 dB
Reverse Soft Handoff Gain	0.00	Min. FM Interference	-118 dBm
		Minimum Reverse Traffic Eb/No	6 dB
		Reverse Interference Ratio	60%
		Target Soft Handoff	85%
		Number of Paging Channels	1
		Misc. Forward Link Gain	0 dB
		Misc. Reverse Link Gain	-8.4 dB
		Orthogonality Factor	-10 dB
		Sectorization Efficiency	0.85
		Calculated Distance	RH (Radio Horiz.)
		Cutoff Signal	-110 dBm
		Diffraction Coefficient	1

* To view these default parameter values, follow these steps: (i) highlight CDMA scenario in the [Scenario List] window, (ii) click [properties...] in Edit menu to bring up [Change Scenario] dialog box, (iii) click on [CDMA Data...], and then the [CDMA Scenario Data] dialog box, in which the default values are stored, will be opened.

Index

A

access channels, 47-66
    acknowledgement, 53-54
    attempts, 47, 50, *51, 52*
    average persistence delay, 55-56, *56, 57*
    calculation of capacity, 64-66
    capacity, 58-66
    delays
        persistence test, 54-58
        quadrature spreading, 53
    failed attempts, 54
    initiation of probe sequence, 54
    interference *vs.* arrival rate, 62, *62*
    message capsule, 50
    messages, attempts, 50, *52*
    parameters, table of, 60
    performance, *60, 61, 62*
    persistence test, 52-58
    PN randomization, 53
    power level, probes, 53
    power used, 59-61
    preamble, 50
    probes, 47, *50,* 50-53
    probe sequences, 52-54
    protocol, 47-54, *48-49*
    psist(n), 54
    response messages, 50, *51*
    simulation model, 59-62
    slots, 50, 53, 59-61
    SMS messages, 64-65
    successful attempts, 58
    traffic assumptions, 63-64

Page numbers in italics indicate illustrations.

utilization rates, 64-65, *65*
active set, 85, 95-99
air path loss, 119-22
ANSI J-STD-008, 1-2
    persistence test, 54
    pilot signals, 23
    soft handoff parameters, 97-101
antennas, 175-203
    base station receiver, 110-14
    beamwidth, 176, 178
    collocated RF stations, 188-203
    cross-correlation coefficient, 183, 185-87
    directional, 176-77, *179*
    directivity, 176-77
    diversity, 175, 183-87, *184*
    downtilt, 177, *181*, 181-83
    efficiency, 176-77
    fading, 175
    gain, 176-81
        polarization diversity, 187
        space diversity, 184
    half-power beamwidth, 176
    height, 177, 180-82, 184
    horizontal spacing, 183
    input impedance, 177
    isolation guidelines, 188-203
        adjacent TX/RX bands, 189
        criteria, 193-95
        degradation, 189
        desensitization, 189-90
        IMP3, 193-94, 199-203
        IMP interference, 189-93
        intermodulation product interference, 190-92
        receiver compression point, 193
        receiver desensitization, 194
        receiver overload, 189, 192-93, 195
        RX filter and receiver, 188-90, *189*
        SAI, 193-95
        total power, 193
        TX amplifier and filter, 188-90, *189*
    main lobe, 176
    multiple collocated RF stations, 188-203
    omnidirectional, 176-78, *178*
    orthogonally polarized, 175, 184-87
    polarization, 177
    polarization diversity, 184-87, *185*, *186*
    propagation paths, 179
    radiation patterns, 176, 178
    radiation properties, 176
    radio propagation geometry, *180*
    SAI (safe antenna isolation), 193-95
    sector channel capacity, 179
    sectorization efficiency, 179
    sectorization gain, 178
    side lobe, 176
    signal cross-correlation, 183
    space diversity, 175, 183-84, *184*, 187
    three-sector configurations, 177-78, *178*
    tilted transmitting, 184, 187

vertical spacing, 183-84
attenuation, base station signal, 118

B

baseband filters, 24
base stations
    antennas (*See* antennas)
    blocking probability, 166-74
    capacity, 127
    CEs *vs.* user per sector, table of, 172
    distance between, 116
    filter, 24
    forward-link budgets, 122-25
    forward links, 9-10
    interference, soft handoffs, 94
    mobile station interference, *134*
    noise, 129
    noise *vs.* cell loading, *134*
    pilot pseudo-noise sequences, 1
    pilot signal reuse distances, 29-41
    power calculations, table of, 123
    power requirements, 9-10
    propagation loss, table of, 123
    received power, 93-94, *94*
    receiver total interfering noise, 112
    signal attenuation, 118
    time shifts between pilot signals, 28
beacon cells, 107
beamwidth, 176
BHCA, 75, *75*
bit energy, forward-link budget, 124

blocking probability of sectored cells, 166-74
border area, 105-6
border cells, 102-3
BWAF (Bandwidth Adjustment Factor), 190

C

call-dragging, 87
CAMs (channel assignment messages), 74-75, *75*
candidate set, 85, 98-99
capacity, 2-3, 8-11, 127-37
    average required signal/user, 132
    base station noise, 129
    cell loading interference, 133
    channel pooling, 137
    co-channel interference, 129-30
    cost per additional mobile, 132-33
    dynamic, 9
    E_b/I_o, 130, 132-35
    Erlang, 3, 165
    forward-link, 133-37
    in-cell interference, 129
    interference, 127
    pilot power allocation, 135
    pole point, 131-32
    power allocation, 135-36
    power pole, 131
    power requests, 135
    receiver interference margin/channels, 131

reverse-link, 128-33
softness, 132
system performance, 136-37
voice activity, 11
voice quality, 135
carrier to thermal noise power ratio, 161
CDMA (Code Division Multiple Access)
    attributes, 8-11
    Channel Numbers, Band Class 1, 14
    defined, 7
    forward links, 9-10
    PCS common air interface standard, 1
    power control, 10
    reverse links, 8-9
    soft handoff, 11
    voice activity, 11
cells
    boundaries, 87, *155-56*
    CEs per sector, 165
    coverage, 139-41, 143
    distance between, 116
    distance relationship between, *162*
    impinging traffic characteristics, 166
    loading, *134*
    number assignment rule, *44*
    transition zone, 103
cellular CDMA common air interface standard, 1
CEs (channel elements), 3
    additional for soft handoffs, 165
    capacity dependent on, 165
    Erlang capacity, *171, 172*
    per sector, 165
    pooling, 165
    traffic, requirements, 165-74
    users per sector, table of, 172
channels
    access (*See* access, channel)
    allocation availability table, 16
    assignment messages, 66
    availability, 15-16
    frequency validity, 16
    minimal spacing, 15
    numbering, 14-15
    preferred, 16-17
    requests (*See* access attempts)
    scanned, 17
    spacing, 14
    TMDA, dividing with, 19-21
chips, conversion to miles, 30
co-channel interference, 129-30, 135, 177
collocated RF stations, 188-203, *189*
coverage, 139-64
    antenna characteristics, 180
    antenna downtilt, 182
    cell loading, 139-41, 143
    distance equation, 141, *142*
    external interference, 141
    forward links, 143-45
    hard handoff area coverage probability, *148*
    hard handoff E_c/I_o threshold, *146-47*
    hexagonal layout, *146*
    location coverage probabilities, 145
    numerical evaluation, 145
    pilot detection, 143-45

Index   243

pilot E_c/I_o threshold, *149*
pilot power allocation *vs.* coverage, *157-59*
power limitations, 143
probability at location, 164
probability for pilot channel, derivation, 160-64
reverse links, 139-42
signal strength for N mobiles, 140
SIR requirements, 140
soft handoffs, 143-45
three-way soft handoff, E_c/I_o, *153-56*
trade-offs, 145
two-way soft handoff, *150-53*
values used in computing, table of, 142
Criterion R1 model, *38*
Criterion S1 timing diagram, *31, 32*
Criterion S2 timing diagram, *32,* 32-33
cross-polarization discrimination, 186

D

delay spread, 97
demodulating forward links, 28
directivity, 176-77
disjoint multi-carrier areas, 101, 104-7
distance
    antenna propagation, 180
    relationship between cells, *162*
diversity antenna systems, 183-87
diversity gain, 112-14

_DONE message, 69, 75-76
downlink (*See* reverse link)

E

E_b/I_o (bit energy to total noise ratio)
    equation, 112
    forward-link budgets, 122
    forward-link capacity, 133-35
    pole point, 132
    reverse-link budgets, 112
    reverse-link capacity, 129-30
E_c/I_o, pilots, 143, 161-62
efficiency, antennas, 176-77
EIRP (effective isotropic radiated power), 109
electronic serial number (*See* ESN)
equations
    access channel capacity, 64
    antenna downtilt, 181-82
    antenna isolation, 195
    average persistence delay, 55
    average required signal/user, 132
    baseband signals, 25
    carrier to thermal noise power ratio, 161
    channel assignment message occupancy, 75
    coverage probability for pilot channel, 160-64
    delay budget, pilots, 96-97
    distance of antenna propagation, 180
    distance of coverage, 141

duplicate offset pilot signals, 31
E_b/I_o, 112, 130
EIRP, 109
fractional power allocation, 135
general page message occupancy, 72
IMP3, 191
interference, collocated RF stations, 190
isolation, overload, 192
marginal probability, unequal Erlang load, 169
maximum air path loss, 119
mobile station signal measured, 26
mobile to base attenuation, 118
net path loss, 35
outage probability, 117
outage, three-way handoff, 118
overhead page message occupancy, 73
path loss, 115
phase offset, 25
pilot
    coverage probability, 144
    PN sequences, 24
    set selection, 85
    signal reuse distance, 36-38
    signals received, 31
    signals transmitted, 30
polarization cross-correlation coefficient, 185
power required at receiver, 112
R1 criterion, 36-38
R2 criterion, 38-39
receiver energy-per-bit, 111
receiver total interfering noise, 112
required signal strength for N mobiles, 140
sectorization antenna gain, 178-79
sectors marginal probability, 167-69
SMS occupancy, 79
traffic load in Erlangs, 166
two-sector joint probability matrix, 168
VMSs occupancy, 77
Erlang
    capacity, 3, 165, *171*, 172
    load, equal, 168
    traffic, access channel assumptions, 63-64
ESN (electronic serial number), 53
explicit diversity gain, 112-14

F

fading
    antennas, 175
    diversity antenna systems, 183-87
    margin, 114-17
FDMA (frequency division multiple access), 7
FER (frame error rate), 110, 122
forward links, 9-10
    budgets, 122-25
    capacity (*See* capacity, forward-link)
    code channels, 24
    coverage, 143-45
    demodulating, 28

Index

E_b/I_o, budgets, 122
    pilot signals, 24
    polarization power loss, 187
    power control, 10
    spectrum, 13-15
frame offset change, 83-84
frequencies, channel, 14
frequency planning, 17-18
frequency validity, 16
fringe areas, 11

G

gain, antenna, 176-77
general page message (*See* page message)
growth, system, 18
guard zones, 20

H

hand-down, 101-2, 104-6
handoffs, 2, 83-107
    hard (*See* hard handoffs)
    inter-carrier (*See* inter-carrier handoffs)
    populations, 88-91, *91*
    soft (*See* soft handoffs)
    softer (*See* softer handoffs)
handover, 101, 104, 106-7
hard handoffs, 11
    area coverage probability, *148-49*

    coverage, 143-45
    E_c/I_o threshold, coverage, *146-47*
hot spot areas, 84

I

IMP3, 191-94, 199-203
IMP interference, 189-93
initialization state, 54
initiating calls, 8-9
inter-carrier handoffs, 101-7
    area classification, 101
    beacon cells, 107
    border area, 105-6
    border cells, 102-3
    disjoint multi-carrier areas, 101, 104-7
    hand-downs, 101-2, 104-6
    handovers, 101, 104, 106-7
    pocketed multi-carrier areas, 101-4, *102*
    PSMMs, 102, 105-7
    reversing course mobiles, 105
    transition cells, 103
interference
    access traffic, 59-63
    adjacent base stations, 161
    base received by mobile, 160
    capacity, 127
    cell loading, 133
    co-channel, 86-87, 129, 135
        reducing with downtilt, 181-82

reduction by antennas, 177
collocated RF stations, 188-93, *189*
collocation criteria, 193
coverage, relation to, 139
external, 133, 141
IMP, 189-93, 200-203
local, 17-18
non-PCS, 21-22
PCS systems, 15
phase offset reuse, 34-39
pilot signals, 28-30, *30*
power calculations, table of, 124
reverse-link, 111-12
self-generated, 127
severe, 22
SIR (signal-to-interference) ratio, 127
soft handoffs, 92-94
TDMA, 14
intermodulation product interference, 190-92
I pilot PN sequences (*See* pilot signals, I and Q PN sequences)
IS-95
    access channel capacity, 59-66
    link budgets, 109-25
    reverse link, *110*
IS97, 1
IS98, 1
IS-95A, 1, 23, 66-80
IS-95B
    message types, 70
    paging channel, 80
isolation, 188-203

BWAF (Bandwidth Adjustment Factor), 190
criteria, 193-95
IMP3, 191, 193-94, 199-203
intermodulation product interference, 190-92
receiver
    compression point, 193
    desensitization, 194
    noise floor, 193
    overload, 192-93, 195
RX filter and receiver, 188-90, *189*
RX sensitivity degradation, 193
SAI (safe antenna isolation), 193-95
TX amplifier and filter, 188-90, *189*

L

link budgets, 2, 109-25
    forward-link budgets, 122-25
    reverse (*See* reverse-link budgets)
location coverage probability, *146-47*

M

minimum performance standards, 1
mobile EIRP, 109
mobile stations
    access parameters, 47-66
    access protocol, 47-54, *48-49*
    apparent range, 53

Index

call processing, 26-28
capacity, 127
coverage, 139-42
handoffs, 2
idle state, 69
initialization state, 26-27, 54
non-slotted mode, 69
persistence test, 52-53
phase offset reuse interference, 34
pilot signal search, 26-27
power requirements, 8
received power from base antenna, 180
reverse-link budgets, 109-22
signal measured equation, 26
slotted mode, 69
system determination substate, 54
tilted transmitting antennas, 184, 187
traffic channel state, 27-28
transmitter strength, 91-92, 141
waking up, 69
MSC (mobile switching centers) hard handoffs, 84
multi-carrier use (*See* inter-carrier hand-offs)

N

neighbor set, 85, *96*, 96-97, 99
net path loss equation, 35
NHBR_MAX_AGE, 99

O

offsets (*See* pilot signals, phase offsets)
omnidirectional cells
    antennas, 178
    handoff populations, *91*
operation band, 192, 199
order messages, 74-75, *75*
overhead message, 66
overhead page message occupancy, 72-74, *74*

P

page messages, 66, 72, *73*
paging channel, 66-69
    assumptions, table of, 71
    capacity, 66-80
    capsules, 67
    channel assignment message, 66, 74-75, *75*
    data burst message occupancy, 79-80
    data rate, 67
    _DONE message occupancy, 75-76
    general page message occupancy, 72, *73*
    IS-95B, 70, 80
    message lengths, table of, 71
    message names, table of, 70
    message types, 69-71
    order message occupancy, 74-75, *75*
    overhead message, 66, 72-74, *74*

page message, 66
SCI bit, 67
slots, 67, 69, 73-74
SMS occupancy, 79-80
structure, 67-69, *68*
synchronization, 67
total occupancy, 80
VMSs occupancy, 76-78, *78*
path loss, 114-16, *115*
    forward-link budget, 125
    maximum supportable, 119-22
    mobile station coverage, 139
    street-level, 121
PCS Spectrum Allocation table, 13
PCS systems, interference from, 15
persistence test, 52-53
    ANSI J-STD-008, 54
    average persistence delay, 55-56, *56, 57*
    delays due to, 54-58
    psist(n), 54
    threshold (P), table of, 55
phase assignment problem (*See* pilot signals, phase assignment)
phase offsets, 1, 23, 32-45
PILOT_INC, 27-34
    Criterion R1, 35-38, *38*
    Criterion R2, 38-39
    Criterion R3, 39-41, *40*
    lower limit, 29-34
    recommended value, 42
    upper limit, 41
pilot signals, 9

acquisition by mobile stations, 26-27
active set, 95-96
ANSI J-STD-008, 23
assignment, 23-45
baseband, 25
cell number assignment rule, *44*
cell transmittals equations, 30
channels, 3
coverage, 143-45, 164
criterion
    R1, 35-38, *38*
    R2, 38-39
    R3, 39-41, *40*
    S1 and S2, 31-33, *31-32*
delay budget, 96-98
delay spread, 97
derivation of coverage probability, 160-64
duplicate offset equations, 31
E_c/I_o, 161-62
E_c/I_o threshold, *149*
energy per chip at mobile, 160
equations of elements, 24
forward links, 24
generation, *24*
I and Q PN sequences, 23-24
I generation from baseband, *26*
incorrect, 28
interfering, 28-29
IS-95A, 23
mobile station call processing, 26-28
mutual interference, 28-34, *30*
net path loss equation, 35

phase assignment, 23-45
    cell numbering, 44-45
    plan, *43*
    procedure, 41-45
    reserve for growth, 42-43
phase measurements, 27-28
phase offsets
    available, 42
    equation, 25
    lower bound, 32
    reuse, 34-45, *35*
PILOT_INC
    lower limit table, 34
    recommended value, 42
    system parameter, 27-34
    upper limit, 41
PN sequences, 1, 25, 32-33
power percentage allocation, 145, *148, 157-59*
propagation delays, 95-97
PSMMs, 28, 102, 105-7
reference, 94-95
reuse distances, 28-34, 36-38
search windows, 94-98
sector indistinguishability, 39-41, *40*
separation criteria, 28-34
set selection equation, 85
set transactions, *86*
signal received equations, 31
strongest, 163
synchronization, 25
T_ADD threshold, 27
undesired fingers, 29

Walsh function, 24
pilot strength measurement messages (*See* PSMMs)
ping-pong effect, avoiding, 117
PN3383, 1
PN (pseudo noise) sequences, 1, 25, 32-33
PN randomization, 53
pocketed multi-carrier areas, 101-4, *104*
polarization, 177
pole points, 9, 131-32
power
    base station, 9-10
    control, 4, 10
    cost per additional mobile, 132-33
    mobile station, 8
    poles, 9, 131
preferred channels, 16-17
    table of, 17
    TMDA interference, 19
propagation delays, 95-97
pseudo-noise (PN) sequences (*See* PN (pseudo noise) sequences)
psist(n), 54
PSMMs (pilot strength measurement messages), 28, 102, 105-7

Q

Q pilot PN sequences (*See* pilot signals, I and Q PN sequences)
quadrature spreading delays, 53

R

radiation pattern, antennas, 176
Rake receivers, 28
ratio of bit energy to interference (*See* E_b/I_o)
receivers
    antenna, base station, 110-14
    compression point, 193
    desensitization, 189-90, 194
    interference margin, channels, table of, 131
    noise floor, 189, 193
    operation band, 192
    overload, 189, 192-93, 195
    sensitivity, 111
remaining set, 85, *96,* 96-97, 99
reuse distance equations, 36-38
reverse-link budgets
    air path loss, 119-22
    base station receivers, 110-14
    E_b/I_o, 112
    EIRP, 109
    explicit diversity gain, 112-14
    fading margin, 114-17
    FER (frame error rate), 110
    impacts of changes, 119-21
    interference, 111
    in-vehicle penalties, 121
    losses, 109
    maximum air path loss equation, 119
    minimum noise floor, 111
    multiple paths, 112
    outage probability, 116-17
    outage, three-way handoff, 118
    path loss, 114-16, *115*
    performance, *113*
    receiver interference margin, 121
    receiver sensitivity, 111
    sample budget table, 120
    shadow fading, 114, 116-17
    shadowing variable, 118
    soft handoff gain, 117-19
    street-level path loss, 121
reverse-link capacity (*See* capacity, reverse-link)
reverse links, 8-9, *110*
    access channel, 58
    access traffic interference, 59-63
    coverage, 139-42
    power control, 10
    spectrum, 13-15
rule for cell number assignment, *44*
RX filter and receiver, 188-90, *189*
RX sensitivity degradation, 193

S

SAI (safe antenna isolation), 193-195
SCI (synchronized capsule indicator) bit, 67, 75-76
search windows, 94-101
sector pilot (*See* pilot signals)
sectors
    antennas, 175

Index

border, 102
distinguishing, assignment, 23 (*See also* pilot signals)
geographical separation of pilot signals, 34
handoff populations, *91*
phase offsets, 34-42
subscribers supported, table of, 63
user limits, 165
users, probability of N, 166
shadow fading, 114, 116-17
Short Messages Service (*See* SMS)
signal loss, polarization, 185
SIR (signal-to-interference) ratio, 127
slotted mode, 69
slow fading, 114
SMS (Short Messages Service)
    access channel capacity, 64-65
    data burst message occupancy, 79-80
    VMS, method used for, 77-78
softer handoffs, 84-101 (*See also* soft handoffs)
soft handoffs, 11, 84-101
    additional CEs for, 165
    base received power, 93-94, *94*
    cell boundary *vs.* E_c/I_o, *153*
    coverage, 143-45
        contour, 88-89
        vs. power, *151-52*
    delay budget, 96-98
    delay spread, 97
    E_c/I_o threshold, *150-51*
    gain, reverse link, 117-19

inherent gain, 91-93
interference, 92-94
NHBR_MAX_AGE, 99
omnidirectional cells, *91*
parameters, 97-101
performance, 87-97
pilot set selection equation, 85
pilot signals, 84-85
pocketed multi-carrier areas, 101-4, *102*
population model, 88-91, *91*
propagation delays, 95-97
reference pilots, 94-95
search windows, 94-97, 99-101
sectored cells, *91*
spectrum coordination, 18
SRCH_WIN parameters, 99-101
T_COMP, 99
T_DROP, 88, 98-100
three-way, 92, *119, 153-55*
transmitter strength, 91-92
two cell *vs.* three cell, 92
two-way, *150-52*
vs. analog handoffs, 86-87
Spectrum Allocation (PCS) table, 13
spectrum coordination, 13-22
    allocation, 14-15
    channel availability, 15-16
    channel numbering, 14-15
    guard zones, 20
    inter-system issues, 18-22
    intra-system issues, 17-18
    minimal spacing, 15

non-PCS interference, 21-22
PCS Spectrum Allocation, 13
preferred channels, 16-17
TMDA, dividing with, 19-21
spreading sequences, 23
SRCH_WIN parameters, 95-97, 99-101
street-level path loss, 121
subscribers supported per sector, table of, 63
system interference, 10

T

T_ADD, 27, 98
T_COMP, 99
TDMA (time division multiple access), 7
    channels, 14
    dividing spectrum with, 19-21
    interference from, 14, 19-21
T_DROP, 98-100
thermal noise, forward-link budget, 124
timing diagram for Criterion S1, *31, 32*
timing diagram for Criterion S2, 32-33, *32*
traffic engineering, 165-73
    blocking probability, equal Erlang load, 168
    cell characterization, 166
    Erlang load, equal, 168
    K-sector case, 170-73
    joint probability matrix, 168, 170
    load in Erlangs, 166
    marginal probability, unequal Erlang load, 169
    three-sector case, 169-70
    two-sector case, 166-69
transition cells, 103
transition zone, 103
trunking efficiency, 165
two-sector traffic case, 166-69
TX amplifier and filter, 188-90, *189*

U

undesired fingers, 29, 35-39
uniform multi-carrier areas, 101
uplink (*See* forward link)

V

VMI (voice mail indication) message, 76
VMSs paging channel occupancy, 76-78, *78*
voice activity capacity, 11

W

Walsh function, 24

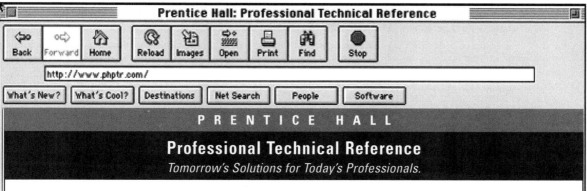

Keep Up-to-Date with
PH PTR Online!

We strive to stay on the cutting-edge of what's happening in professional computer science and engineering. Here's a bit of what you'll find when you stop by **www.phptr.com**:

@ Special interest areas offering our latest books, book series, software, features of the month, related links and other useful information to help you get the job done.

Deals, deals, deals! Come to our promotions section for the latest bargains offered to you exclusively from our retailers.

$ Need to find a bookstore? Chances are, there's a bookseller near you that carries a broad selection of PTR titles. Locate a Magnet bookstore near you at www.phptr.com.

! What's New at PH PTR? We don't just publish books for the professional community, we're a part of it. Check out our convention schedule, join an author chat, get the latest reviews and press releases on topics of interest to you.

Subscribe Today! Join PH PTR's monthly email newsletter!

Want to be kept up-to-date on your area of interest? Choose a targeted category on our website, and we'll keep you informed of the latest PH PTR products, author events, reviews and conferences in your interest area.

Visit our mailroom to subscribe today! **http://www.phptr.com/mail_lists**

LICENSE AGREEMENT AND LIMITED WARRANTY

READ THE FOLLOWING TERMS AND CONDITIONS CAREFULLY BEFORE OPENING THIS SOFTWARE MEDIA PACKAGE. THIS LEGAL DOCUMENT IS AN AGREEMENT BETWEEN YOU AND PRENTICE-HALL, INC. (THE "COMPANY"). BY OPENING THIS SEALED SOFTWARE MEDIA PACKAGE, YOU ARE AGREEING TO BE BOUND BY THESE TERMS AND CONDITIONS. IF YOU DO NOT AGREE WITH THESE TERMS AND CONDITIONS, DO NOT OPEN THE SOFTWARE MEDIA PACKAGE. PROMPTLY RETURN THE UNOPENED SOFTWARE MEDIA PACKAGE AND ALL ACCOMPANYING ITEMS TO THE PLACE YOU OBTAINED THEM FOR A FULL REFUND OF ANY SUMS YOU HAVE PAID.

1. **GRANT OF LICENSE:** In consideration of your payment of the license fee, which is part of the price you paid for this product, and your agreement to abide by the terms and conditions of this Agreement, the Company grants to you a nonexclusive right to use and display the copy of the enclosed software program (hereinafter the "SOFTWARE") on a single computer (i.e., with a single CPU) at a single location so long as you comply with the terms of this Agreement. The Company reserves all rights not expressly granted to you under this Agreement.

2. **OWNERSHIP OF SOFTWARE:** You own only the magnetic or physical media (the enclosed software media) on which the SOFTWARE is recorded or fixed, but the Company retains all the rights, title, and ownership to the SOFTWARE recorded on the original software media copy(ies) and all subsequent copies of the SOFTWARE, regardless of the form or media on which the original or other copies may exist. This license is not a sale of the original SOFTWARE or any copy to you.

3. **COPY RESTRICTIONS:** This SOFTWARE and the accompanying printed materials and user manual (the "Documentation") are the subject of copyright. You may not copy the Documentation or the SOFTWARE, except that you may make a single copy of the SOFTWARE for backup or archival purposes only. You may be held legally responsible for any copying or copyright infringement which is caused or encouraged by your failure to abide by the terms of this restriction.

4. **USE RESTRICTIONS:** You may not network the SOFTWARE or otherwise use it on more than one computer or computer terminal at the same time. You may physically transfer the SOFTWARE from one computer to another provided that the SOFTWARE is used on only one computer at a time. You may not distribute copies of the SOFTWARE or Documentation to others. You may not reverse engineer, disassemble, decompile, modify, adapt, translate, or create derivative works based on the SOFTWARE or the Documentation without the prior written consent of the Company.

5. **TRANSFER RESTRICTIONS:** The enclosed SOFTWARE is licensed only to you and may not be transferred to any one else without the prior written consent of the Company. Any unauthorized transfer of the SOFTWARE shall result in the immediate termination of this Agreement.

6. **TERMINATION:** This license is effective until terminated. This license will terminate automatically without notice from the Company and become null and void if you fail to comply with any provisions or limitations of this license. Upon termination, you shall destroy the Documentation and all copies of the SOFTWARE. All provisions of this Agreement as to warranties, limitation of liability, remedies or damages, and our ownership rights shall survive termination.

7. **MISCELLANEOUS:** This Agreement shall be construed in accordance with the laws of the United States of America and the State of New York and shall benefit the Company, its affiliates, and assignees.

8. **LIMITED WARRANTY AND DISCLAIMER OF WARRANTY:** The Company warrants that the SOFTWARE, when properly used in accordance with the Documentation, will operate in substantial conformity with the description of the SOFTWARE set forth in the Documentation. The Company does not warrant that the SOFTWARE will meet your requirements or that the operation of the SOFTWARE will be uninterrupted or error-free. The Company warrants that the media on which the SOFTWARE is delivered shall be free from defects in materials and workmanship under normal use for a period of thirty (30) days from the date of your purchase. Your only remedy and the Company's only obligation under these limited warranties is, at the Company's option, return of the warranted item for a refund of any amounts paid by you or replacement of the item. Any replacement of SOFTWARE or media under the warranties shall not extend the original warranty period. The limited warranty set forth above shall not apply to any SOFTWARE which the Company determines in good faith has been subject to misuse, neglect, improper installation, repair, alteration, or dam-

age by you. EXCEPT FOR THE EXPRESSED WARRANTIES SET FORTH ABOVE, THE COMPANY DISCLAIMS ALL WARRANTIES, EXPRESS OR IMPLIED, INCLUDING WITHOUT LIMITATION, THE IMPLIED WARRANTIES OF MERCHANTABILITY AND FITNESS FOR A PARTICULAR PURPOSE. EXCEPT FOR THE EXPRESS WARRANTY SET FORTH ABOVE, THE COMPANY DOES NOT WARRANT, GUARANTEE, OR MAKE ANY REPRESENTATION REGARDING THE USE OR THE RESULTS OF THE USE OF THE SOFTWARE IN TERMS OF ITS CORRECTNESS, ACCURACY, RELIABILITY, CURRENTNESS, OR OTHERWISE.

IN NO EVENT, SHALL THE COMPANY OR ITS EMPLOYEES, AGENTS, SUPPLIERS, OR CONTRACTORS BE LIABLE FOR ANY INCIDENTAL, INDIRECT, SPECIAL, OR CONSEQUENTIAL DAMAGES ARISING OUT OF OR IN CONNECTION WITH THE LICENSE GRANTED UNDER THIS AGREEMENT, OR FOR LOSS OF USE, LOSS OF DATA, LOSS OF INCOME OR PROFIT, OR OTHER LOSSES, SUSTAINED AS A RESULT OF INJURY TO ANY PERSON, OR LOSS OF OR DAMAGE TO PROPERTY, OR CLAIMS OF THIRD PARTIES, EVEN IF THE COMPANY OR AN AUTHORIZED REPRESENTATIVE OF THE COMPANY HAS BEEN ADVISED OF THE POSSIBILITY OF SUCH DAMAGES. IN NO EVENT SHALL LIABILITY OF THE COMPANY FOR DAMAGES WITH RESPECT TO THE SOFTWARE EXCEED THE AMOUNTS ACTUALLY PAID BY YOU, IF ANY, FOR THE SOFTWARE.

SOME JURISDICTIONS DO NOT ALLOW THE LIMITATION OF IMPLIED WARRANTIES OR LIABILITY FOR INCIDENTAL, INDIRECT, SPECIAL, OR CONSEQUENTIAL DAMAGES, SO THE ABOVE LIMITATIONS MAY NOT ALWAYS APPLY. THE WARRANTIES IN THIS AGREEMENT GIVE YOU SPECIFIC LEGAL RIGHTS AND YOU MAY ALSO HAVE OTHER RIGHTS WHICH VARY IN ACCORDANCE WITH LOCAL LAW.

ACKNOWLEDGMENT

YOU ACKNOWLEDGE THAT YOU HAVE READ THIS AGREEMENT, UNDERSTAND IT, AND AGREE TO BE BOUND BY ITS TERMS AND CONDITIONS. YOU ALSO AGREE THAT THIS AGREEMENT IS THE COMPLETE AND EXCLUSIVE STATEMENT OF THE AGREEMENT BETWEEN YOU AND THE COMPANY AND SUPERSEDES ALL PROPOSALS OR PRIOR AGREEMENTS, ORAL, OR WRITTEN, AND ANY OTHER COMMUNICATIONS BETWEEN YOU AND THE COMPANY OR ANY REPRESENTATIVE OF THE COMPANY RELATING TO THE SUBJECT MATTER OF THIS AGREEMENT.

Should you have any questions concerning this Agreement or if you wish to contact the Company for any reason, please contact in writing at the address below.

Robin Short
Prentice Hall PTR
One Lake Street
Upper Saddle River, New Jersey 07458

ABOUT THE CD-ROM

Welcome to the *Handbook of CDMA System Design, Engineering, and Optimization* CD. The software on this CD requires Windows 95 or Windows NT.

This CD contains the demonstration version of Cellular Engineering 4 (CE4) referred to in this book. A detailed description of CE4 and the procedure for installing the program can be found in Appendix C of this book.

Technical Support

Prentice Hall does not offer support for this software. If there is a problem with the media, however, you may obtain a replacement CD-ROM by emailing a description of the problem to

disc_exchange@prenhall.com